The International Library of Ethics, Law and Technology

Volume 15

Series Editors
Anthony Mark Cutter, Lancashire, UK
Bert Gordijn, Ethics Institute, Dublin City University, Ireland
Gary E. Marchant, Center for the Study of Law, Science, and Technology,
Arizona State University, USA
Colleen Murphy, University of Illinois at Urbana-Champaign, Urbana, IL, USA
Alain Pompidou, European Patent Office, Munich, Germany
Sabine Roeser, Dept. Philosophy, Delft University of Technology, Delft,
The Netherlands

Editorial Board
Dieter Birnbacher, Institute of Philosophy, Heinrich-Heine-Universität, Germany
Roger Brownsword, King's College London, UK
Ruth Chadwick, ESRC Centre for Economic & Social Aspects
of Genomics, Cardiff, UK
Paul Stephen Dempsey, Institute of Air & Space Law, Université
de Montréal, Canada
Michael Froomkin, University of Miami Law School, FL, USA
Serge Gutwirth, Vrije Universiteit, Brussels, Belgium
Henk ten Have, Duquesne University, Pittsburgh, USA
Søren Holm, University of Manchester, UK
George Khushf, Center for Bioethics, University of South Carolina, USA
Justice Michael Kirby, High Court of Australia, Canberra, Australia
Bartha Maria Knoppers, Université de Montréal, Canada
David Krieger, The Waging Peace Foundation, CA, USA
Graeme Laurie, AHRC Centre for Intellectual Property and Technology Law, UK
René Oosterlinck, European Space Agency, Paris
Edmund Pellegrino, Kennedy Institute of Ethics, Georgetown University, USA
John Weckert, School of Information Studies, Charles Sturt University, Australia

More information about this series at http://www.springer.com/series/7761

Federica Lucivero

Ethical Assessments of Emerging Technologies

Appraising the moral plausibility
of technological visions

Federica Lucivero
Department of Social Science, Health and Medicine,
  Faculty of Social Science and Public Policy
King's College London
London, UK

ISSN 1875-0044   ISSN 1875-0036   (electronic)
The International Library of Ethics, Law and Technology
ISBN 978-3-319-23281-2   ISBN 978-3-319-23282-9   (eBook)
DOI 10.1007/978-3-319-23282-9

Library of Congress Control Number: 2015949923

Springer Cham Heidelberg New York Dordrecht London
© Springer International Publishing Switzerland 2016
This work is subject to copyright. All rights are reserved by the Publisher, whether the whole or part of the material is concerned, specifically the rights of translation, reprinting, reuse of illustrations, recitation, broadcasting, reproduction on microfilms or in any other physical way, and transmission or information storage and retrieval, electronic adaptation, computer software, or by similar or dissimilar methodology now known or hereafter developed.
The use of general descriptive names, registered names, trademarks, service marks, etc. in this publication does not imply, even in the absence of a specific statement, that such names are exempt from the relevant protective laws and regulations and therefore free for general use.
The publisher, the authors and the editors are safe to assume that the advice and information in this book are believed to be true and accurate at the date of publication. Neither the publisher nor the authors or the editors give a warranty, express or implied, with respect to the material contained herein or for any errors or omissions that may have been made.

Printed on acid-free paper

Springer International Publishing AG Switzerland is part of Springer Science+Business Media (www.springer.com)

It is not enough to change the world. That happens anyway and generally beyond our control. What matters is to interpret this change, specifically in order to lead it. So that this world does not change further outside of ourselves, ultimately becoming a world-without-us.

(G. Anders. Die Antiquiertheid des Menschen, vol 2)

Preface

New File. The blank page is scary. Open recent > Notes for Chapter 1. Ctrl + A, Ctrl + C. Better to start with this. Now, Ctrl + V on the blank page. This paragraph is a good start for my book. Ctrl + X, Ctrl + V, select paragraph, move up. Uhm, no… Ctrl + Z. A tag pops up in the bottom right corner of my screen: "This is your rest break. Make sure you stand up and walk away from your computer on a regular basis. Just walk around for a few minutes, stretch and relax." I can check my Facebook page in my break. Or perhaps I shouldn't. I should install the software that limits my access to Facebook during working hours. This is killing my productivity! A walk might be better. Oh, wait: Ctrl + S.

These lines portray a typical moment in the experience of my daily life as I wrote this book. Yet 33 years ago, to the average academic, they might have seemed as though they were emerging from the pages of a science fiction story. It was in 1982, the year I was born, that WordPerfect Corporation introduced WordPerfect 1.0, destined to "become one of the computer markets most popular word processing programs".[1] With new technologies, such as this, come new innovations: novel tools become available, different skills are required, old abilities become superfluous, new problems emerge, whilst previous problems are redefined and addressed with original technical solutions from which novel moral obligations arise, together with needs and desires. This is just one example among the many possibilities showing how the introduction of technologies deeply affects our daily practices by altering our knowledge, habits, perceptions, capabilities, and values.

Would it have been worthwhile to reflect on the impacts of computing and word processing on writing practices 30 years ago? Would such a reflection have affected the development of new hardware and software to avoid the occurrence of Repetitive Strain Injury syndrome? Would it have impacted policy makers and managers to grasp the sudden changes of writing and working practices? Would such a reflection on potential impacts have helped parents to better understand their children? Finally, would such a prospective thinking on a future practice even be possible at all? As the Italians say: "history cannot be done with 'if' and 'but'". That is, retrospective speculations on how things could have been different 30 years ago are not purposeful.

---

[1] See http://www.computerhope.com/history/1982.htm.

Rather, what can be done is a prospective investigation of the relevance of such reflection on current emerging technologies.

The importance of a reflection on the desirability of emerging technologies has been addressed in the policy tradition of Technology Assessment (hereafter TA). In the early days, TA offices would produce reports that would guide policy makers' decisions concerning new science and technologies. Experts in science, technology and economics were considered the best candidates for the task of producing such reports on technologies' impacts. Later, the argument was made that, if technology plays such a big role in citizens' lives, everyone in society should participate in decision making about new technologies. Thus, not only the experts, but also the citizens should have a say in deliberating on the desirability of emerging technologies. If emerging technologies should be democratically evaluated, then the values and understanding of desirability should be clarified and openly discussed.

Etymologically, the word "assess" comes from the Latin *ad-sedēre*, meaning to sit, referring to the sitting position of judges comparing and estimating the "value of (property or income) for the purpose of apportioning its share of taxation".[2] An assessment is an act of determining an amount (for example properties or income) and estimating (or comparing) its value with respect to a quality standard (for example spending or purchasing power). In this sense, assessing emerging technologies is an evaluative activity: it does not simply describe what impact a technology might have, it also suggests whether this impact is good or bad according to some "value". Although Technology Assessment activities are always evaluative of the desirability of technologies, the meaning of "desirability" has been interpreted in a variety of ways. A technology may have desirable consequences when it enhances the economy of a country, when it improves people's health, or the environment or when it eases people's everyday lives. Different economic, scientific, social or moral values can be mobilized to assess the desirable impacts of a new technology. We can define this evaluation as normative when a technology is assessed with respect to explicit norms or authoritative standards. Such standards may be legal or moral norms. The adjective "normative" is often used to qualify a judgment in opposition to a "descriptive" account. While the latter aims at describing a state of affairs, in a presumably objective and value-neutral manner, the former is a judgment, an evaluation based on some previously established values. As it will be discussed throughout this book, this distinction has been criticized by a broad scholarship in the humanities and social sciences, which has argued that facts are always value-laden and descriptions are never neutral accounts of facts, but always framed in a way that promotes some aspects and marginalize and exclude others. Despite such agreement on the normative character of any account, there has been a long tradition of TA exercises that has not directly engaged in discussions concerning the implied, sometimes hidden, values that guided such assessments. Discussions and debates on the goodness and rightness of new science and technologies on the basis of moral norms and principles have, instead, been relegated to the realm of ethics that traditionally deals with

---

[2] "assess, v.". OED Online. September 2011. Oxford University Press. http://www.oed.com/view/Entry/11849?redirectedFrom=assess (accessed October 06, 2011).

controversies concerning moral values. As Chap. 1 will show, this has been acknowledged as a weakness of current TA approaches and accordingly, some attempts have been made to overcome such limitations and include spaces for the discussion of moral values in assessments of emerging technologies. These attempts have aimed at shedding light on the normative character of decisions concerning these technologies and, in some cases, critically discuss their appropriateness. Falling into this tradition, this book aims at investigating ways to do "ethical" assessments of emerging technologies, that is assessments that disclose the normative nature of visions and decisions about emerging technologies, by exploring their moral purport.

In moving towards this goal, this book focuses on one specific aspect: the fact that in these assessments we focus on science and technology that is still emerging. Prospective evaluation is not easy. In our daily lives, we have a hard time anticipating the consequences of our own actions. The task becomes even harder when the consequences depend on a large network of interacting players. The greatest challenge comes when we want to evaluate the desirability of these future consequences. Should we then give up with the attempt of meaningful discussions about the potential role that future technologies may play in our lives? Some policy analysts, sociologists and philosophers have argued that this is not ideal given that the expectations of the ways in which emerging technologies will change our lives, the promises of their benefits as well as the threats of potential losses determine our present decisions. Visions of technologies guide our decision-making processes; they justify our choices and exclude alternatives. This happens on multiple levels: when politicians deliberate on investing public money for the research and development of new technologies; when healthcare managers decide how to re-organize the system for efficiency; when researchers select the focus of their research; when entrepreneurs consider what to invest in; when adolescents decide on a course of study; when patients exclude certain treatments but accept others; when doctors empower their patients in the decision-making process. Reflecting on the meaning of emerging technologies enables our society to understand current technological developments and their role in our practices in the very near-future. This understanding allows us to interpret them and hopefully to make more cognizant decisions in the present.

This book contributes to the debate of "how" the desirability of emerging technologies can be assessed. In particular, it addresses the question of how to deal with "expectations" on emerging technologies when assessing their desirability. Emerging technologies are, by definition, "not there yet" and we can only assess their desirability by looking at the current expectations of their future development. Yet, these expectations do not provide stable grounds for philosophers and ethicists to ask moral questions about the desirability of emerging technologies. Why? The grammar of expectations clarifies this point. If I say that I expect to finish writing my book in a few months, my expectation communicates that I believe and I hope that I will finish my book in a few months. In the act of expecting, there is an element of belief that something can happen, for example that I have enough chapters written. There is also an element of interest that something should happen, for example that I want to finish my book. Furthermore, I can utter this expectation in order to convince my editor to be ready to receive my book. Since my expectation depends on my beliefs and interests, and can have a specific function, it cannot be taken as a starting point from which the value of my book can be assessed. The same line of reasoning applies to expectations that emerging technologies will produce some societal benefit.

Philosophers of technology and applied ethicists cannot take expectations surrounding emerging technologies as descriptions of states of affairs. In the case of technologies that are still emerging, the normative assessment of the desirability of emerging technologies has to start by appraising the quality of expectations surrounding those emerging technologies. How can the epistemological robustness of such expectations be assessed in view of a normative reflection on their desirability? This methodological question is the central focus of this study. The goal of this book is to articulate, implement and justify the approach to assessing the plausibility of expectations surrounding new and emerging science and technologies. This book argues that ethical assessments of emerging technologies are always plural and context specific. Although the two technologies taken as case studies are both examples of emerging screening innovation, the proposed methodology can be used for different types of technologies.

This book is organized in three parts: Part I presents the problems, research questions and the approach that is taken in order to address them; Part II describes and justifies the three steps of the proposed approach, through an exploration of the case of an emerging technology for cancer screening, the "Nanopil"; Part III, addresses the possibilities for applying and implementing the three-step approach described in Part II. In Chap. 1, introduce the general debate on the assessment of emerging technologies. I focus on a gap between two traditions used to assess technologies, namely Technology Assessment and institutional ethics. The former tradition fails to deal with questions about the desirability of emerging technologies, while the latter lacks a sociological sense of the context. Different approaches have addressed this gap, but the aspect of epistemological uncertainty that characterizes emerging technologies seems to remain understudied. Chapter 2 expands on the topic of "expectations" and the need to assess their quality. In this chapter, I present a body of literature that justifies the need to develop a methodology for assessing the quality of expectations. I first turn to the literature on the sociology of expectations to investigate their social construction. According to the literature, expectations should not be taken at face value, because they have a strategic and performative role. The literature on "visions" emphasizes that it is indeed important to assess the desirability of the values and norms implied in visions of future technologies. Since this normative content is not always explicit, it should be disclosed before it can be assessed. These analyses of expectations are enlightened by the literature on empirical philosophy of technology that points out that technologies often do much more, and very different things, than they were originally intended to. Consequently, I argue that, before asking whether these implicit norms and values are desirable, one should check how plausible it is that they will indeed be realized. To address this question, I develop an analytic and methodological approach which I refer to as "plausibility assessment". This approach is based on a three-step process that requires the articulation of three elements of these expectations: the expected artifact, its potential use and the anticipated valuable impacts.

Such an analytical framework is further described in the second part of this book where the expectations of a specific emerging technology are used as an exemplary case: the "Nanopil", an ingestible device for in vivo screening of intestinal cancer. Chapter 3 illustrates how to address analyzing expectations about a future artifact. After introducing public expectations surrounding the Nanopil, I explain why further analysis is needed and how it can be done. Then, I present my research design and the analysis of expectations of the Nanopil, explaining how this analysis helps to address the question of the plausibility of expectations. Chapter 4 addresses the question of how to analyze expectations of the potential use of an emerging technology. Using the example of the Nanopil, I explain why they need to be assessed and what conceptual and methodological tools help with this. These preparatory analyses set the stage for addressing the main question pertaining to the plausibility of visions in Chap. 5. In this chapter, I return to the question of how plausible it is that certain values and desirable worlds will indeed be realized by a new technology. The plausibility of the expectations of the Nanopil is assessed on three levels: how likely is it that the artifact will promote the expected values? To what extent are these values desirable? And how likely is it that a technology will instrumentally bring about a desirable consequence?

The third part of this book discusses how the three-step approach developed in Chaps. 4, 5 and 6 can be applied to other cases and used to develop tools for integrating ethical inquiry in TA exercises. In Chap. 6, I apply this analytical and methodological framework to another technology: the Immunosignatures. At the end of this chapter, I discuss those parts of my approach that have been adjusted in order to analyze this specific technology, and those parts of the analysis that remain the same. Chapter 7 shows how a plausibility assessment can improve the debate on the desirability of emerging technologies. Using the pragmatist normative framework, this chapter explains how democratic deliberations can be improved by triggering stakeholders' moral imagination through scenarios and vignettes. The analysis of two pilot workshops, organized with the scientists and engineers developing the Nanopil and the Immunosignatures, highlights the opportunities and limits associated with these tools. Chapter 8 returns to the discussion outlined in Chap. 1. It discusses the contribution of the proposed approach to assessing the expectations of plausibility to the fields of applied ethics and Technology Assessment. This final chapter aims at explaining how this study contributes to the goal of ethically assessing emerging technologies by improving the conditions for democratic deliberation on the desirability of emerging technologies.

London, UK   Federica Lucivero

Acknowledgements

Although I will never manage to thank adequately all those who have, directly or indirectly, played an important role in the realization of this book, I will nevertheless make an attempt. I would like to thank Marianne Boenink and Tsjalling Swierstra for their continuous and caring intellectual and emotional support in the drafting phase. I am grateful to Philip Brey, Frans Brom, Erik Fisher, Alfred Nordmann, Arie Rip, and Peter Paul Verbeek, and two anonymous reviewers for their constructive and critical comments. Many friends and colleagues have read and commented on these chapters – Aimee, Anna Laura, Eleni, Lieke, Lise, Pierre, Ronald – thanks for your constructive critiques and encouragements. A special thanks goes to Clare and Lucie for their thorough comments on the last revised chapters.

My research would be all 'head' and no 'body' without the availability and generosity of the people I interviewed in conducting my case studies. In particular, I wish to thank the BIOS group at Twente University and the Centre for Innovation in Medicine at Arizona State University. These groups are working on the two emerging technologies that I used as case studies for this research project. I wish to thank all the members of these groups, especially those who have spent long hours discussing their research with me. The expertise, patience and enthusiasm that they showed in our conversations and through their participation in the activities that I organized have been crucial for this whole study.

I have been part of great academic communities that have inspired the content of this book and encouraged me to publish it: my colleagues at the Department of Philosophy at the University of Twente, the Tilburg Institute for Law Technology and Society, and the Social Science Health and Medicine, especially Barbara Prainsack, for her support in the very last phases prior to publication. Also, I would like to thank Sally Eales for carefully editing several chapters of this book. Of course, I take full responsibility for every mistake or imperfection still present in the text. Finally, I am grateful to those who helped with the material realization of this book: Chris Wilby, my editor, who supported me with great patience and encouragement and the 3TU Centre for Ethics and Technology and the Socio-Technical Integration Research (STIR) who have financially supported this project.

Much gratitude and love goes to my family, especially to my dad for his affectionate availability and prompt support when I was lost with layouts and formats. And Alessio, who supported me with patient love and encouragement while I was working on this publication, and who both inspires and grounds me every day.

Contents

Part I

1 Democratic Appraisals of Future Technologies: Integrating Ethics in Technology Assessment ... 3
- 1.1 Appraising Emerging Technologies ... 4
- 1.2 From the Myth to the History: The Evolving Social Mandate of Technology Assessment ... 5
- 1.3 "Institutional" Ethics of Technology ... 11
  - 1.3.1 Ethical Bodies and the Regulation of Biomedical Research ... 12
  - 1.3.2 Normative Evaluation of Emerging Technologies and Advice to Policy-Makers ... 14
  - 1.3.3 Outsourcing Ethical Reflection ... 15
- 1.4 Limitations in Traditions Assessing Technologies ... 16
  - 1.4.1 The Normative Deficit in TA ... 17
  - 1.4.2 The Technological and Sociological Deficit in Institutional Ethics ... 21
- 1.5 The Need for Integrating Ethical Inquiry in TA ... 23
- 1.6 Between Grounding and Exploring: The Contribution of This Study ... 29
- References ... 31

2 Promises, Expectations and Visions: On Appraising the Plausibility of Socio-Technical Futures ... 37
- 2.1 Expecting Future Science and Technologies ... 37
- 2.2 The Social Construction of the Future ... 41
- 2.3 The Guiding Normativity in Technological Visions ... 43
- 2.4 Beyond an Instrumentalist View: Technology and Morality ... 47

	2.5	Analyzing Expectations' *Plausibility*: A Proposal	51
		2.5.1 Desirability Versus Plausibility	52
		2.5.2 Breaking Down the Plausibility Question...............................	53
		2.5.3 In Search of Plausibility..	55
		2.5.4 Three Strategies to Appraise Plausible Visions......................	56
	References...		59

Part II

3	**The Mechanism in the Pill: From Abstract Images to Detailed Descriptions**	65
	3.1 Visions of Promising Technologies: The Nanopil	65
	3.2 Promises of Emerging Artifacts..	67
	3.3 Rhetoric and Black-Boxes	68
	3.4 A Note on Methods	71
	3.5 The Nanopil: Tales of an Emerging Object	73
	3.5.1 From an Idea to a Project	73
	3.5.2 An Idealized System and its Building-Blocks	75
	3.5.3 The Functional Components and Their Material Conditions	78
	3.6 From the Lab "Details" Back to the Big Picture	80
	Appendix...	82
	References...	83

4	**The Doctor in the Pill: From "Technical" Details to Social Practices**..	85
	4.1 Expectations of Artifacts in Use	85
	4.2 (Fictive) Scripts and Actor-Worlds	86
	4.3 Research Design	90
	4.4 The Nanopill: Tales of an Emerging Practice	92
	4.4.1 Nanopil Designers-World	92
	4.4.2 Comparing Actors' Worlds: Current Screening Practice and Future Trends	94
	4.4.3 Users' Preference and Resistance	97
	4.5 Conclusions	99
	References...	100

5	**The Good in the Pill. Assessing the Plausibility of Visions of Desirable Worlds**	103
	5.1 Visions of Desirable Worlds	103
	5.2 Different Expected Artifacts and Different Values	105
	5.3 Plurality of Values Among Actors	109

	5.4	Impacts of Technologies and the Moral Landscape	111
		5.4.1 Mediation	113
		5.4.2 The Co-production of Technology and Morality	117
	5.5	Conclusion	118
	References		120

Part III

6 Expecting Diagnostics, Diagnosing Expectations. The Plausibility Framework in Use ... 125
  6.1 Immunosignatures and the Healthcare Revolution ... 125
  6.2 Research Design ... 128
  6.3 Immunosignatures: A "Simple" Concept ... 130
    6.3.1 Reconstructing the History of the Concept ... 130
    6.3.2 Concepts and Components in Research Practice ... 132
    6.3.3 Some Conditions for Immunosignatures to Work ... 134
  6.4 The Expected Context of Use ... 135
    6.4.1 The Many Applications of Immunosignatures ... 136
    6.4.2 Assessing and Enriching Fictive Scripts from Situated Perspectives ... 139
  6.5 Immunosignatures and a Desirable World ... 142
    6.5.1 Articulating Moral Connotations in Different Technological Platforms ... 143
    6.5.2 Stakeholders and Normative Divergence ... 145
    6.5.3 The Interactions Between Immunosignatures and Morality ... 147
  6.6 Discussion ... 149
  Appendix ... 151
  References ... 153

7 Scenarios as "Grounded Explorations". Designing Tools for Discussing the Desirability of Emerging Technologies ... 155
  7.1 In Search of a Normative Framework ... 156
    7.1.1 Democratic Deliberation as a Normative Ideal: A Pragmatist Approach ... 157
    7.1.2 Triggering Moral Imagination ... 160
  7.2 Scenarios as Tools to Foster Moral Imagination ... 161
  7.3 Plausible Scenarios for "Grounded Explorations" ... 164
  7.4 Techno-Moral Vignettes and Scenarios in Action ... 165
    7.4.1 Workshop 1: Immunosignatures ... 166
    7.4.2 Workshop 2: the Nanopil ... 172
  7.5 Discussion ... 176

      Appendix: Techno-Ethical Scenarios and Techno-Moral Vignettes 179
           Techno-Ethical Scenarios on Immunosignatures 179
           Techno-Moral Vignettes on Nanopil... 186
      References... 188

8 Building-Blocks for Ethical Assessments of Emerging Technologies.. 191
      8.1 Between "Grounding" and "Exploring".. 191
      8.2 Towards Ethical Assessments of Emerging Technologies................. 194
      8.3 Ethical Expertise? Interpreting and Intervening................................ 198
      8.4 Open Questions 199
      References... 201

Abbreviations

ANT	Actor-Network Theory
CIM	Center for Innovations in Medicine
CTA	Constructive Technology Assessment
EGE	European Group on Ethics in Science and New Technologies
ELSI/A	Ethical Legal and Social Issues/Aspects
eTA	Ethical Technology Assessment
FOBT	Fecal Occult Blood Test
HGP	Human Genome Project
ICT	Information and Communication Technologies
ImSg	Immunosignatures
NEST	New and Emerging Science and Technology
NP	Nanopil
OTA	Office of Technology Assessment
pTA	Participatory Technology Assessment
R&D	Research and Development
RRI	Responsible Research and Innovation
S&T	Science and Technology
STS	Science and Technology Studies
TA	Technology Assessment
VA	Vision Assessment

Part I

Chapter 1
Democratic Appraisals of Future Technologies: Integrating Ethics in Technology Assessment

> *O most ingenious Theuth, the parent or inventor of an art is not always the best judge of the utility or inutility of his own inventions to the users of them (Plato, Phaedrus)*

Abstract The mandate to the assessment of new technologies has been evolving for the last four decades according to societal and political contexts. As such, this chapter explains evolving trends towards more participatory and deliberative models of Technology Assessment (hereafter TA) and increasingly broader sets of aspects (beyond efficiency and health impacts) of ethical inquiry. It discusses in which sense TA initiatives have been criticized for a normative deficit, while bioethics councils and applied ethics approaches to the study of new technologies have been accused of a technological and sociological deficit. In addressing the question of how to integrate ethical inquiry in TA and how to account for societal contexts, the literature has focused on the importance of accounting for techno-social co-shaping and stakeholders' conflicts while exploring the moral dimensions, framing and values inherent in new technologies or actors' controversies. Within this enterprise, the issues of emergence, uncertainty and dynamic evolution that characterize the technologies under investigation deserve attention. The debate on "speculative ethics" is introduced as a reflection on the possibilities of knowledge concerning technologies that are still emerging and, as such, do not fully exist yet. If emerging technologies do not yet exist and we can only address them as prospective projections, how can we guarantee that an assessment of their desirability is epistemologically robust? The contribution of this book lies in addressing this question.

Keywords Technology assessment • CTA • Ethics • Emerging technologies • Normative deficit

1.1 Appraising Emerging Technologies

In the Platonic dialogue *Phaedrus,* Socrates tells the myth of Theuth. Theuth – the god "inventor of many arts, such as arithmetic and calculation and geometry and astronomy and draughts and dice, but his great discovery was the use of letters" – talked in front of Thamus, the king of Egypt, and

> […] showed his inventions, desiring that the other Egyptians might be allowed to have the benefit of them; he enumerated them, and Thamus enquired about their several uses, and praised some of them and censured others, as he approved or disapproved of them. It would take a long time to repeat all that Thamus said to Theuth in praise or blame of the various arts. But when they came to letters, This, said Theuth, will make the Egyptians wiser and give them better memories; it is a specific both for the memory and for the wit. Thamus replied: O most ingenious Theuth, the parent or inventor of an art is not always the best judge of the utility or inutility of his own inventions to the users of them. And in this instance, you who are the father of letters, from a paternal love of your own children have been led to attribute to them a quality which they cannot have; for this discovery of yours will create forgetfulness in the learners' souls, because they will not use their memories; they will trust to the external written characters and not remember of themselves. The specific which you have discovered is an aid not to memory, but to reminiscence, and you give your disciples not truth, but only the semblance of truth; they will be hearers of many things and will have learned nothing; they will appear to be omniscient and will generally know nothing; they will be tiresome company, having the show of wisdom without the reality. (Plato, *Phaedrus*)

This myth offers a narrative introduction to the topic of this book: the prospective appraisal of the quality of emerging technologies. Thamus is a scrupulous governor, carefully assessing the praise and blame of new inventions, and extensively approving and disapproving of them. Thamus legitimizes his appraisal of Theuth's inventions through his ability to detect qualities that would otherwise remain hidden, due to the enthusiastic nature of the inventor. The technology under examination, in this case the alphabet, is new and emerging, never been seen or used by any Egyptian before. The distinction between the role of the evaluator and inventor is clear-cut: on the one side there is Theuth enthusiastic about how his technologies will benefit Egyptians, on the opposite side there is Thamus weighing the possible detrimental impacts on society. This attempt to assess emerging technologies *avant la lettre* defines who is legitimated to appraise the social desirability of new technologies and, as such, assigns to the "king" a broader, more comprehensive view than the inventor. This chapter discusses the evolving social mandate around the assessment of new technologies. It explains evolving trends towards more participatory and deliberative models for assessing technologies (beyond Plato's "king model") and increasingly broader sets of aspects (beyond efficiency and health impacts) of ethical inquiry (Sects. 1.2 and 1.3). It discusses in which sense initiatives of technology assessment have been criticized for a normative deficit, while bioethics councils and applied ethics approaches to the study of new technologies have been accused of a technological and sociological deficit (Sect. 1.4). In addressing the question of how to integrate ethical inquiry in TA and how to account for societal contexts, the literature has focused on the importance of accounting for techno-social co-shaping and stakeholders' conflicts while exploring the moral dimensions, framing and values inherent in new technologies or actors' controversies. Within this enterprise, the issues of emergence, uncertainty and dynamic evolution that characterize the technologies under investigation, deserve attention. The debate on "speculative ethics" is introduced as a reflection concerning the possibilities of knowledge concerning technologies that are still emerging and, as such, do not exist yet (Sect. 1.5). If emerging technologies do not yet exist and we can only address them as prospective projections, how can we guarantee that an assessment of their desirability is epistemologically robust? The contribution of this book lies in addressing this question (Sect. 1.6).

1.2 From the Myth to the History: The Evolving Social Mandate of Technology Assessment

The notion that society should assess the desirability of technologies is fairly recent. Even in the aftermath of WW2, when the images of the atomic bomb in the Japanese skies were still vivid in people's minds, the contribution of scientific research to societal progress was still widely acknowledged. The faith in the *endless frontier* of scientific research, whose progress would inevitably reward society with goods, used to guide investments in basic research in the US (Bush 1945). Society trusted scientists who worked according to a mandate in order to contribute to social progress. According to the influential work of sociologist Robert Merton (1973), science as a social institution has a normative structure and is self-regulated by an intrinsic *ethos*.[1] If in science there is an intrinsic control mechanism that regulates its development and the impact of research outcomes, there is no need for interfering or steering scientific development from the outside.

Only as recently as the 1960s – while advancements in scientific research and new technologies were a key component of Cold War public narratives[2] – did the increasing fear of technologies' potential negative consequences begin to surface in policy and media discourse. Such a fear resulted in a destabilization of faith in the

---

[1] In particular, four moral norms of behavior guide the appropriate scientific practice: universalism, communism, disinterestedness and organized skepticism (Merton 1973). Science serves the social function of providing certified knowledge since scientists conform to the four norms and provide society with sincere and accurate information about the given world and future forecasts. Within this view, scientific knowledge is, on the one hand, objective and neutral with respect to interests and values and on the other hand, science is intrinsically guided by ethics with respect to the four norms.

[2] In 1957, the Soviet Union launched *Sputnik 1*, the first artificial satellite to be put into the Earth's orbit. This event, initiating the so-called "Space Age", triggered many reactions of the western scientific community contributing to the shift away from an "endless frontier" model in science and technology governance.

internal control mechanism within the scientific community.³ Assessing the technological impact on society became a major task for policy makers.⁴ This new attitude is reminiscent of Thamus in Plato's myth: since technology developers are not fit to assess the potential benefits and damages of their inventions, institutions should take over this task. Considered as a management tool, the general aim of Technology Assessment (hereafter TA) was to reduce the costs of technologies' detrimental effects by anticipating potential impacts. Such TA activities were expected to help governments and parliaments to mobilize the most appropriate financial, political and regulatory resources in the governance of technologies. For this purpose, in 1972 the Office of Technology Assessment (OTA) was established to support US Congress in dealing with cutting edge science and technology in many fields such as medicine, telecommunications, agriculture, materials, transportation, and military defense.

In these early days, TA was an instrument for policy analysis to inform Congress and to orient strategic decision-making. The expected output of such assessments was a factual and neutral expert-based report. The OTA's mandate was rescinded in 1995. By then, however, the institutionalization of TA had already arrived in Europe. In the late 1980s and early 1990s several TA organizations were established in many European states and at the supranational level of the European Union: the Office Parlamentaire d'Evaluation des Choix Scientfique et Technologique (OPECST) in France, the Parliamentary Office of Science and Technology (POST) in UK, the Technikfolgenabschaetzungsbuero Deutscher Bunsdestag (TAB) in Germany, the Danish Technology Board (DTB), the Netherlands Office for Technology Assessment (NOTA),⁵ and the Scientific and Technological Options Assessment (STOA) at the European Union level. The European counterparts of OTA presented a different approach from the American institution. As Smits et al. explain:

> TA was viewed in a broader and more sophisticated way, not simply avoiding negative effects but pursuing a better integration of science and technology in society (2008: 7)

The European TA offices were not exclusively geared to produce a robust, factual and objective report. Some offices in particular, like the Danish Board of Technology and the Dutch NOTA, building on ideas of participatory democracy,

---

[3] Emblematic in this respect is the prominent role of Aerospatiale research and war technologies in the filmography of the later 1960s (see for example Stanley Kubrick's *2001: a Space Odysseys* and *Dr. Strangelove*). In these imaginary narratives, new technologies, supposed to celebrate the evolution of the human species and its progress, turn against human beings in a chaotic and uncontrolled way.

[4] Gibbons and colleagues (1994) describe a trend that emerged in the second half of the twentieth century wherein scientific research in context-driven (carried out in the context of application) and directed towards solving a problem across disciplinary boundaries. This transdisciplinary "mode 2" knowledge production (see also Nowotny et al. 2001) or post-normal science (Funtowicz and Ravetz 1993) has gone along with the need for a societal and political assessment of scientific and technological production.

[5] After being evaluated in 1993, the mission of the NOTA is readjusted, emphasizing the role of the organization to support decision-making and societal debate. To mark this shift, since 1994, the name of the office changes to the Rathenau Institute.

1.2 From the Myth to the History: The Evolving Social Mandate of Technology...

adopted interactive and participatory methods. A growing number of stakeholders were involved in different interactive activities such as awareness initiatives, consensus conferences, scenario workshops, citizen hearings, and deliberative mappings, among others (Klüver et al. 2000).

There have been several attempts to define TA, distinguishing typologies (see for example Abels 2007), identifying modes (van Est and Brom 2012), reviewing methods (Tran and Daim 2008; Decker and Ladikas 2004) and reflecting on the theory (Grunwald 2009) of TA. In Van Est and Brom's words (2012): "it is not easy to get a hold on TA. TA is rather hard to define. It is not a separate field of scientific research, nor is it a well-defined, clear-cut practice." The authors[6] continue by reviewing existing TA approaches and identifying four modes of understanding and performing TA: (1) classical, (2) participatory, (3) argumentative and (4) constructive. Each of these modes is inspired by different scientific disciplines, performs different democratic functions and is institutionalized differently.

The "classical TA" model is exemplified in the OTA approach. Inspired by expert-based policy analysis, this type of TA had the role of informing decision-makers in a representative democratic model and was institutionalized in a parliamentary office (*ibidem*). Smits and Leyten (1991) use the effective metaphor of "watchdog TA": technology assessment was established with a centralized "top-down" approach that would provide a systematic early warning evaluation of potential impacts of new technologies. In this model, technology is conceptualized as a given, a static autonomous entity which has direct impacts on society. Such impacts can be traced prospectively by a group of experts and controlled by governments. The unwanted impacts of technology were mostly conceived in terms of risk for health or environment. For example: the harmful influence of gas and toxic metal emissions on environment; or alternatively the negative social and psychological consequences of increasing computerization and automation in the workplace.

When it arrived in Europe, the goal of TA shifted from one of early warning to one of providing options for policy development.[7] Instead of focusing on the report as a "product" of TA, the "process" of TA is considered as valuable in itself (van Eijndhoven 1997). Often inspired by deliberative democracy theories (van Est and Brom 2012), participatory trends in Technology Assessment have been brought forward as a way of broadening the decision making process in S&T policies through the involvement of a variety of stakeholders in the political debate. In some cases, like Denmark, participatory TA brings the decision-making process from citizens' representatives in the Parliament, to the citizens themselves. To do this, citizens participate in "consensus conferences". Promoters of "participatory TA" (pTA) expected that enlarging stakeholder participation in TA activities would have a two-fold goal: it contributes towards making governance more participatory and expertise more democratic. This means that governance of emerging technologies

---

[6] Rinie van Est and Frans Brom are respectively research coordinator and head of the Technology Assessment division at the Rathenau Institute.

[7] Although van Eijndhoven (1997) points out that this goal was already embraced in the last years of the OTA.

becomes more democratic (and therefore better) by involving in the decisional process those who will be affected by S&T policies, and by including additional sources of knowledge and expertise (besides the technical and scientific ones) in the process of assessing technologies (Joss and Bellucci 2002).[8] In this sense, TA progressively endorses a deliberative democratic model. According to deliberative democratic theories, an institution's policies are legitimized by their accountability, that is the justification and articulation of public policy by giving reasons for it (Dryzek 2000; Gutmann and Thompson 2002). In this "talk-centric democratic theory" (Chambers 2003), debates and discussions are expected to make the decision-making process more democratic and pluralistic. They create the conditions for participants to produce reasonable and well-informed decisions in which they take into account the positions of the other participants.

In addition to these parliamentary TA institutions, other types of TA communities have emerged: industrial, academic and executive[9] communities have developed a variety of TA concepts in the European landscape (Smits et al. 1995). In general, these approaches cluster a number of bottom-up activities done *by* and *for* different actors involved in the development, management and usage of new technologies. TA was not embodied in one institution, but was multiform and decentralized. Smits and Leyten characterize this generation of TA with another metaphor: the "tracker [dog] TA" (1991). This metaphor captures the idea that, rather than a one-time assessment, TA became a process of ongoing dialogue supporting actors' decision-making process. From the early days of TA, an increasing number of participants and perspectives were included in the process of assessing technologies. In addition, the type of issues or impacts that qualified as important for assessment was broadened. If, at the dawn of TA, only health risks and economic impacts were assessed, later on "broader societal concerns" became central. Van Est and Brom (2012) highlight how some forms of TA can be defined as "argumentative" because they are based on the understanding of policy discourse analysis, ethics, and sociology of S&T and aim at deepening the political debate bringing into it core values and visions of different actors.

Whereas classic institutional TA aimed at producing reports in which new technologies were reviewed and their potential direct and indirect consequences were described, this second generation, or "paradigm", of TA (van Eijndhoven 1997) drifts away from the notion that TA has to contribute towards forecasting possible impacts and produce some objective knowledge. Implicit in the early TA approaches was the assumption that a value-free description of facts (about technologies and consequences) could be provided in order to guide political action. However, soon it became clear that forecasting the future effects of technology was difficult and controlling them almost impossible. The Collingridge dilemma (1980) has often been mobilized recently in order to point out the impasse in an attempt to control technological development at an earlier rather than later stage. At this stage

---

[8] This was in line with the shift from an expert-based model to a participatory model in public policy (Fischer and Forester 1993).

[9] The executive TA communities are exemplified by non-governmental organizations.

technology is easier to direct, but uncertainties are higher and effects more unpredictable. The uncertainties and challenges in anticipating effects of emerging technologies are even higher if a non-linear perspective on innovation is considered. As shown by innovation studies and social constructivist approaches, technologies do not affect society according to a linear logic of cause and effect (Latour 1987; Callon et al. 1986; Pinch and Bijker 1984). Instead, society and technologies shape or co-produce one another and co-evolve (see for example Jasanoff 2004). This perspective has been taken up within the scholarly community of sociologists working on TA and in particular in the tradition of "Constructive Technology Assessment" (CTA) (Rip et al. 1995; Schot and Rip 1997). This approach aims at broadening the development process – rather than only the political decision-making process – involving stakeholder and citizens (van Est and Brom 2012).

CTA have stimulated a long-lasting scholarly reflection on the roles, theories and methods of technology assessment. Originating in the mid-1980s in the Netherlands, CTA has a long theoretical and applied history that has inspired more recent approaches such as Real-Time TA in the USA (Guston and Sarewitz 2002) and the interactive learning and action approach in EU (Broerse and Bunders 2000). All these approaches aim at a more "symmetrical" version of participatory technology assessment by engaging in an early-stage dialogue with different actors. The application of CTA in a number of projects has contributed to the development of a consistent methodology. Such a methodology consists of "anticipating potential impacts and feeding these insights back into decision making, and into actors' strategies" (Schot and Rip 1997: 251). Rather than aiming at controlling the technological end-product and expecting to steer it in one direction or another, these scholars draw attention to the process of technological innovation as a space and object of assessment.

Studies of technology dynamics provide a fundamental theoretical tool to analyze and identify possibilities to modulate technology development at an early stage (Rip et al. 1995). The key insight is that technology and society mutually influence each other – they co-evolve[10] – therefore their dynamics cannot be studied in isolation. Another key concept is that actors, actions and practices become entangled and at some moment stabilize. This stabilization produces long lasting interactions in which actors and activities become mutually dependent. Structures emerge which enable some actions and constrain others. For example, when technology developers initiate a project they have a certain context of application in mind, say a medical or military application. Therefore they are "allied" with doctors or governments.

---

[10] CTA is inspired by the extensive literature from the fields of Science and Technology Studies (STS) and innovation studies that describe how interests, power-relations, and social structures play a role in scientists and engineers' work and how technology and society mutually shape one another. This STS-oriented TA draws on a range of studies: Bruno Latour's reflections on the social dynamics in the laboratory (1987), Callon (1986) and Law (1986) on the roles of human and non-human "actants" in complex techno-social systems (Actor Network Theory, see also Latour 2005); the social construction of technological artifacts and the co-shaping dynamics or co-production between technology and society (Bijker 1995; Jasanoff 2004); and, research on the creative role of users in technological innovation (von Hippel 1988; Akrich 1992; Oudshoorn and Pinch 2003).

These early alliances close up the possibility for other interactions and lock the development of the technology according to a certain path in which interactions with other stakeholders (say, the food industry) are constrained. In the vocabulary of Innovation Studies, socio-technical configurations are "entrenched" in "lock-ins" that constrain technological developments along specific paths. A sociological analysis of innovation dynamics and stakeholders' interactions points out "emerging irreversibilities" and "path dependencies" at an early stage of technological development, (Robinson and Merkerk 2006; Robinson 2009; Rip and te Kulve 2008). This allows stakeholders and analysts to point out futures that are already present (or "endogenous") in some present configurations and choices and to understand the role that expectations on emerging technologies have in such a process (van Merkerk and van Lente 2005; van Merkerk 2007; te Kulve 2011).

CTA's focus shifts from the "assessment" of emerging technologies typical of early warning TA to the "modulation" of the ongoing process of technology development and stakeholder transactions by providing them with information about techno-social dynamics and patterns. In doing so, CTA scholars and practitioners developed various tools to engage a wide range of stakeholders in reflecting prospectively on diverse aspects of technological innovations. The analysis of patterns of socio-technical dynamics is integrated in the preparation of interactive sessions (CTA-workshops) in which relevant stakeholders are invited to participate. These heterogeneous settings offer a "protected space" to stakeholders to reflect on technological developments, and to position themselves with respect to others. During a CTA workshop, stakeholders are in a deliberative setting in which they have to articulate their position, defend their arguments, criticize and learn from others. This interactive exercise is supposed to enhance stakeholder reflexivity and learning as well as to broaden their perspectives, thus enabling them to play an active and more aware role in the innovation process.

The political and societal mandate of TA increasingly focuses on the process rather than the product of political decision-making in science and technology policies. Between 2004 and 2010, a number of policy reports in the UK,[11] EU and US (Sclove 2010) claimed that emerging science and technology assessments should be a democratic exercise in which users and citizens are offered a say in decision-making on technology and innovation. Broader social participation and more active roles of users also foster an integration of different perspectives into the discussion of emerging technologies. This also broadens the scope of these discussions. In addition to questions concerning health, environmental and safety issues, more uncertain and ambiguous social and ethical issues are brought up.[12] Whereas

---

[11] HM Treasury/Department of Trade and Industry/Department of Education and Skills 2004 *Science and innovation investment framework 2004/2014*. HM Treasury, London (quoted in Kearnes and Wynne 2007).

[12] According to Swierstra (1997), this change in the issues to be discussed is recognizable in the media and in the political discourse on technologies and presents three elements. First, the discourse moves from questions of survival in relation to technology to good-life issues. Second, the classic dichotomy between society, values and culture, on one hand, and technology, facts and instrumental logic, on the other, is abandoned in favor of an idea of "technological culture". Third,

societal aspects of emerging technologies in the past were taken into account only at the end of the innovation process (if at all), they have been increasingly seen as an important ingredient at early stages of innovation. As the next section will show, such a trend is also visible in the institutional mandate and mission of (bio)ethical committees, that were initially created to monitor the conduct of scientific research involving human beings and to progressively broaden the range of their interests in new and emerging technologies as a matter for ethical discussion.

1.3 "Institutional" Ethics of Technology

The growing interest of Western societies and policy makers in the assessment of new science and technology materializes not only in the increasing development of TA activities and institutions, but also in the emergence of several bodies in charge of evaluating the "ethics" of S&T developments. As an academic specialty, ethics is the philosophical discipline that addresses the question of how human beings act in a good or right manner. Consequently, ethics focuses traditionally on the behavior, intentions, and aims of human actions. In every day language, we hear sentences such as "she acted ethically" or "his behavior was unethical", to refer to judgments of a good/bad action on the basis of some moral principles. As opposed to common sense ethics, as a philosophical discipline, ethics as a philosophical discipline includes second-order considerations about morality and individual maxims. It takes into account the foundation of moral theories, definition of concepts, examination and definition of topics and the analysis of methods of reasoning in ethical arguments (Grunwald 2004; Swierstra and Rip 2007). To explain the distinction between morality and ethics, Swierstra and Rip (2007) use the metaphor of the iceberg. Below the sea level, a consistent set of values, norms and standards of good behavior define the landscape of an unquestioned array of principles, models and virtues shared in a certain society. This invisible part of the iceberg, embedded in habits and routines, often undefined and ambiguous, represents morality. The visible tip of the iceberg is ethics, considered as the study of morality. However, ethics and ethical discussion is not only a field of philosophical inquiry. Ethical controversies and debates happen in the "real-world"[13] when actors explicitly discuss, question or defend moral principles and values, often in heated debates or controversies. In this sense, the authors refer to *ethics* as "hot morality" and *morality* as "cold ethics".

---

there is an increasing interest in attributing responsibility in the process of technological development. This trend also appears in the "outsourcing" of ethical evaluation through research funding programs (such as ELSA and RRI) discussed below.

[13] The distinction between ethical debates and "real world" ethics is the starting point of the EU funded project "DEEPEN". Its declared aim is to reach an "integrated understanding of the ethical challenges posed by emerging nanotechnologies in real world circumstances" (see http://www.geography.dur.ac.uk/projects/deepen/Home/tabid/1871/Default.aspx). For an insightful discussion of this topic see Shelley-Egan (2011).

The role of judgments concerning the "good and bad" of a new science or technology has increasingly become more important for political deliberation on science and technology (Swierstra 1997). Societies have progressively acknowledged that technologies should also be regulated according to ethical considerations. Issues concerning values should be part of discussions on new technologies and moral responsibilities in the process of technological development and should be clearly ascribed. In line with these emerging trends several institutions have been established with the role of evaluating new science and technology according to moral criteria.

For analytical reasons, it is possible to distinguish this "institutional" ethics of science and technology with respect to the political mandate or mission that it was assigned. The first mandate is to monitor research practices, draft ethical guidelines and guarantee their respect. A mandate concerns the evaluation of the ethical implications of emerging technologies and advising institutions about desirable policies. A mission consists of the "outsourcing" of ethical evaluation, as in the case of the funding programs for studies of the Ethical Legal and Social Issues (ELSI) – or more recently Responsible Research and Innovation (RRI) – related to new science and technology. These programs fund research into ethical legal and social issues related to emerging technologies that focus on both practices and implications of new technologies. In the following paragraphs these three mandates for institutional ethics will be briefly sketched.

1.3.1 Ethical Bodies and the Regulation of Biomedical Research

A process of "institutionalization" of ethics in scientific research practice can be traced back to the end of the 1940s, when the *Nuremberg Trials* showed the world how Nazi doctors engaged in experiments with human beings in concentration camps against every human right. In 1947, together with the conviction of said doctors, the authorities stated ten principles designed to guide the experimentation on human beings, commonly known as *Nuremberg Code*.[14] Integrated in the *Declaration of Helsinki* these principles include respect for the individual, the right of participants in research to self-determination and the right to make informed decisions concerning participation in clinical research.[15] Furthermore, the declaration defines the role and importance of independent ethical committees (Article 23) that assess the theoretical and practical appropriateness of clinical research protocols and consent procedures. The institutionalization of the ethics of scientific research was

---

[14] The Nuremberg Code is available online at http://www.hhs.gov/ohrp/archive/nurcode.html.

[15] The "Declaration of Helsinki – Ethical Principles for Medical Research Involving Human Subjects", originally issued by the World Medical Association in 1964 was last amended at the 64th WMA General Assembly, Fortaleza, Brazil, October 2013 (available online http://www.wma.net/en/30publications/10policies/b3/index.html).

visible in the establishment of the *National Commission for the Protection of Human Subjects of Biomedical and Behavioral Research* in 1974 as an advisory body to the Department of Health, Education and Welfare. In 1978, the Commission issued the "Belmont Report: Ethical Principles and Guidelines for the Protection of Human Subjects of Research". The report referred to three core principles, that is *beneficence, respect for persons and justice,* whilst also addressing issues such as *informed consent* and *risk-benefit analysis.*

In parallel to the articulation of the normative framework regarding biomedical and clinical research, several research institutes focusing on bioethics were established in the second half of the twentieth century: in 1969 the *Hastings Centre* was established in Garrison (New York) as an independent bioethics research institute with the mission to "address fundamental ethical issues in the areas of health, medicine, and the environment as they affect individuals, communities, and societies".[16] Two years later, the *Kennedy Institute of Ethics* was founded at Georgetown University, which currently hosts one of the most complete libraries with resources on bioethics. Bioethics is commonly considered a discipline that applies theoretical reflection in ethics to concrete moral problems pertaining to biotechnologies. Traditionally linked to the field of medical ethics dealing with the doctor-patient relationships and the virtues of the good doctor, bioethics goes beyond the scope of medical ethics (Kuhse and Singer 2012). The field of bioethics emerged in the 1960s as a reflection on the ethical controversies raised by "revolutionary developments in the biomedical sciences and clinical disciplines" (ibidem: 3) and comprises various disciplines and theories. Since the 1970s a "four-principle approach" by Beauchamp and Childress (2009 [1979]) has dominated bioethical reflection. This approach attempts to include both consequentialist and deontological ethical theories to identify principles and rules that can be applied to particular judgments about cases. Impartial Rule Theory (see for example Clouser 1995), Casuistry (see Albert Jonsen 1986), and Virtue Ethics (see for example, Pellegrino 1995) approaches have also been discussed by bioethicists (see also Beauchamp 1995).

In this context, Western societies have witnessed a formalization and increasing institutionalization of academic disciplines in the field of research ethics, beyond the biomedical field. Bioethical Committees and Commissions offering normative guidelines for scientific research on human beings have proliferated at the national and international level. These bodies' counterparts at the research institute levels are Institutional Review Boards (IRBs) or Research Ethics Committees (RECs) which have the mandate to approve, monitor and review scientific and behavioral research involving humans or parts of human bodies (tissues, embryos, etc). These bodies can either be part of academic institutions and medical facilities or be independent committees. These committees receive proposals of clinical trials that they have to review. They may either accept them, propose amendments or disapprove them. Furthermore the committee is expected to conduct a continuous review of each on-going trial. The driving idea behind the establishment ethical committees is that the practice of doing scientific research

---

[16] The Hasting Centre – Bioethics and Public Policy http://www.thehastingscenter.org/

should be monitored by external bodies rather than being left to scientists. Ethical bodies were however also created in order to advise policy makers about the Research and Development directions based on their ethical implications, in terms of potential benefits and damages, of technological innovation.

1.3.2 Normative Evaluation of Emerging Technologies and Advice to Policy-Makers

As parliamentary offices for TA would provide policy makers with regard for technological innovation, also ethical/ bioethical committees were established in order to provide recommendations concerning the ethical implications of new and emerging technologies. At the international level the Presidential Commission for the Study of Bioethical Issues in United States[17] and the European Group on Ethics in Science and New Technologies to the European Commission (EGE)[18] offer two paradigmatic examples. The EGE is an independent body of 15 experts designated by the European Commission. Its mission is to examine legal and ethical questions raised by emerging technologies and to write reports (or Opinions) that provide the European institution with a preliminary study to prepare and implement European legislation and policies. This transnational (European) independent body was appointed by the European Commission to provide recommendations to institutions on moral conflicts triggered by new technologies.

Established in 1991 as a Group of Advisers on the Ethical Implications of Biotechnology (GAIEB), the European ethics committee was replaced in 1997 by the European Group on Ethics in Science and New Technologies to the European Commission (EGE).[19] The Group soon expanded its sphere of competences beyond the field of biotechnology: in the third and fourth mandate (2000–2005 and 2005–2010) reports were issued on a diverse array of topics, including ICT implants in the human body, agriculture technologies, and nanomedicine. The expansion of the sphere of competences of the EGE embodies the broadening of European institutions' interest for the set of implications of emerging technologies that would go beyond issues of health risks and safety and would include a discussion of cultural, societal, political and ethical matters associated with the technological innovation under examination.

EGE's reports usually comprise an extensive literature review describing the state of the art in a specific field of science and technology, and a sketch of the

---

[17] More information about the Presidential Commission for the Study of Bioethical Issues in United States can be found online at http://bioethics.gov/cms/history

[18] To read about the mandate of the European Group on Ethics in Science and New Technologies to the European Commission, see http://ec.europa.eu/bepa/european-group-ethics/archive-mandates/index_en.htm

[19] The mandate of the GAIEB can be accessed at http://ec.europa.eu/bepa/european-group-ethics/archive-mandates/mandate-1998-2000/gceb_en.htm

relevant legal background along with the relevant ethical and social issues. After this short review, the Opinion is provided in a prescriptive form specifying how European institutions should direct their policies. The legal and ethical principles to which the EGE appeals for justifying its normative position are drawn from the European official documents regarding the protection of fundamental rights (like the Charter of fundamental Rights or the Convention on Human Rights and Biomedicine).[20] In their reports different normative approaches are used to evaluate emerging science and technologies: a "deontological" approach that refers to some prior universally accepted moral principles; a "consequentialist" normative analysis evaluating technology's potential implications for the total welfare; and "virtue ethics" arguments intended to address how technologies affect people's goal of conducting a good life.

1.3.3 Outsourcing Ethical Reflection

Ethical committees are not the only way institutions have promoted ethical reflection on emerging technologies. Specific research programs have been established to fund projects exploring the ethical implications of emerging technologies. Specifically, in the late 1990s the public interest for developments in genetics and medicine contributed to re-adjusting the focus of the institutional and political attention on upstream assessment of ethical and social issues in Research and Development projects. This interest was triggered by the success of the Human Genome Project and the dystopian scenarios of human cloning and genetic enhancement that it fostered.[21] As a result, funding agencies sought to support research on "ethical, legal and social issues/aspects" (ELSI/A) of genetics in North America and Europe.[22]

These programs "outsourced" the promotion of education, the guidance of researchers' conduct and the advice of medical and public policies to scholars in the fields of applied ethics and bioethics, social sciences and law.[23] This idea of integrating ethical, legal and social reflection early on in the innovation process has been developed further by governments at the national level and institutions at the European level. The focus of early ELSI/A projects ranged from issues of privacy

---

[20] "These rights are rooted in the principle of human dignity and shed light on the core European values, such as integrity, autonomy, privacy, equity, fairness, pluralism and solidarity" (European Group on Ethics 2007: 53)

[21] More about the first ELSA-research program established as part of the Human Genome project is available at http://ghr.nlm.nih.gov/handbook/hgp/elsi. The establishment of ELSI programs is further described in Chap. 2 and Sect. 2.1.

[22] See also van Est and Brom 2012 on ELSA projects as academic activities experimenting with different "ways of doing TA upstream alongside techno-scientific research".

[23] Although scenarios towards the institutionalization of ELSA in a "new breed of socio-humanistic consultants" whose need could be stated in codes of conduct or whose role would have a clearer role in decision-making processes have been outlined (Rip 2009).

of genetic information to questions about potential misuses of genetic data (for example in workplaces or schools) or even the impacts of genetic research on concepts of race and humanity. Although the initial focus was on genetics, this ELSI/A trend has spread in other scientific and technological fields, for instance nanotechnology (Fisher et al. 2006), and food technologies (Mepham 2001).

The role of social scientists and humanities scholars has been presented as an integration of societal and ethical reflection in the scientific work that could offer a multidisciplinary perspective. Ethicists and social scientists play the roles of "mediators" or "convergent workers" (Stegmaier 2009; Rip 2009) that foster a process of mutual reflexivity and learning (Calvert Martin 2009). Currently several national and international funding programs have promoted similar types of research under the banner of "Responsible Research and Innovation (RRI)".[24] Although the focus of RRI programs seem to be more focused on interaction with industry than how it was in the case of ELSI programs (Zwart et al. 2014), these research programs and policy discourses have been seen as grounded in both TA and applied ethics traditions (Grunwald 2011; van Est and Brom 2012) as they place emphasis on technology innovation's impacts, responsiveness through anticipation and responsibility towards uncertainties (Owen et al. 2012).[25]

It should be noted that the "ELSI/A" label is often used in general to indicate studies addressing the ethical social and legal aspects of science and technology independently from the funding program. Apart from this, it is nearly impossible to describe a typical ELSI/A approach, as there are many scholars and teams that work on these issues and not all ELSI/A scholars are interested in exploring the ethical aspects of new technologies. In the following I will refer to "ELSI", to focus on those ELSI/A studies that are conducted by ethicists or scholars with an interest in ethical issue related to emerging technologies.

1.4 Limitations in Traditions Assessing Technologies

On the one hand, TA approaches have increasingly broadened the range of impacts to be assessed, from safety, security and environmental risk to broader social impacts including "ethical" issues (related, for example, to good life, social justice or fair access). However, on the other hand, normative assessments of scientific and technological practices have increasingly focused on the assessment of new

---

[24] RRI is presented as the European Commission's approach to the "Science with and for Society" funding program addressing societal and ethical challenges in scientific and innovative research (https://ec.europa.eu/programmes/horizon2020/en/h2020-section/science-and-society). See also the Responsible Innovation Program (MVI) established in the Netherlands since 2008 (http://www.nwo.nl/en/research-and-results/programmes/responsible+innovation)

[25] This brief historical overview on ethics of technology and its modes of institutionalization is of course painted with a very broad brush that does not adequately describe the substantial differences between Europe and the United States – nor the inter-European differences – in the way bioethics and ethics of S&T have been institutionalized.

technologies as an interdisciplinary and collaborative endeavor. Despite this convergent movement, in the last decade several critics have presented the "TA" and "ethics" traditions, approaches and initiatives as being set apart and stigmatized in the original academic disciplines and tradition that have been crucial to their development

Armin Grunwald (1999) explains these "ethics wars"[26] between TA and ethics of technology as a consequence of the hiatus between the academic disciplines that traditionally inspire these approaches: the social sciences and philosophical ethics. In reconstructing this debate in Germany in the 1990s, Grunwald explains how technology assessment has traditionally been addressed as an "operational" endeavor to develop the best approach for engaging stakeholders and for intervening in the process of technological innovation. Ethicists instead, have interpreted the endeavor of assessing technologies as a normative project to assess the moral value of the technology in question. In this analysis, also shared by others (Palm and Hansson 2006; Roelofsen et al. 2008; Brey 2012), TA presents a "normative deficit" while ethical committees and ethicists involved in ELSI studies show a "sociological deficit". TA is appreciated for its attention to democratic processes of decision-making, but presents a "normative deficit" because it is involved too little in the exploration of moral questions and normative discussions on the moral acceptability of the technologies at stake. Ethical committees/ELSI pproaches instead, bring forward these normative and moral questions, but are too little informed by sociological structures and dynamics. Let's look at these critiques more closely.

1.4.1 The Normative Deficit in TA

TA is inherently normative in at least two respects (see Grunwald 2004). First, the idea of *shaping* decision-making processes and technologies for a better societal outcome (Rip et al. 1995) calls for a reflection on the goals and objectives that are desirable for a society and requires an evaluation of the technical options. Second, the design of the TA methods has a normative dimension: the use of terminology, distinction and classification and the very methods to discuss and choose between options are not value neutral and require some normative deliberation.

Classical TA, which aims at producing objective reports that could inform policy makers, presented a normative deficit both in the goal setting and the methods. In fact the OTA reports were considered as objective and a value neutral policy analysis of impacts, and provided information on what *could* be done, rather than what *should* be done (Grunwald 1999). The methodology was informed by expert-based rational policy analysis (van Est and Brom 2012), which entailed a description of

---

[26] Hoeyer (2006) refers to "ethics wars" to indicate the antagonism between two different scholarly traditions involved in ELSI studies: ethics and social science. Not only do these two disciplinary fields rely on different theoretical backgrounds, they also use divergent methodological approaches. As such each of these two have differing attitudes or "ethical codes".

the state of the art, assessment of impacts and consequences and elaboration of alternatives for technology policies. The alternative options and their desirability or acceptability were not evaluated, but left to the judgment of the political system. A spin-off of OTA, the Health Technology Assessment (HTA) defined as a "systematic study of the consequences of the [...] use of technology in a particular context" (Hofmann 2005) was created with a mission of including a broad range of "studies of the ethical and social consequences of technology" (Banta 1997). However, HTA has often been criticized for its narrow focus on the outcomes and costs in modern healthcare systems and for producing only generalizable quantifiable evidence that addresses questions of effectiveness and efficiency of the technology at stake (see Lehoux 2006; Banta and Perry 1997). Although authors seem to agree that ethical and moral issues should be integrated in HTA (Hofmann 2005, 2008; Lehoux and Williams-Jones 2007; ten Have 2004; van Oostwijn et al. 2004), HTA holds a deterministic view that does not account for the relationship between values, society and technology (Clausen and Yoshinaka 2004).

The "normative deficit" of TA, however, emerges not only in traditional TA approaches characterized by expert consultations, but also in more participatory approaches involving different stakeholders. Several authors have criticized participatory TA approaches for their failure to reflect on their impact on policy decisions (Hennen 2012), and inability to justify their democratic basis and thus, legitimacy (Abels 2007). It has also been highlighted that the political and power structures in their methodological set-up are not sufficiently problematized (Blok 2007; Jensen 2005; van Oudheusden 2014). Grunwald (1999) points out that TA focuses on stakeholders' factual acceptance and dodges evaluative exercises on the normative acceptability of technologies. Accordingly, it prohibits the space for "trans-subjectivity" of evaluations on emerging technologies.

> Bargaining takes the factual preferences and values of the concerned parties as valid merely because they are factually given – a kind of naturalistic misconception, which neglects the necessity to argumentatively legitimate actions, and decisions [...] factual acceptance is, indeed not sufficient to allow conclusive decision as to the normative acceptability. (Ibidem: 175)

This means that value-related positions are restricted to the subjective sphere and when normative positions of stakeholders emerge, their acceptability is not assessed. Political philosopher Richard Sclove, in a recent report evaluating the work of the US Office for Technology Assessment (OTA), proposes a quasi-hypothetical example that shows how moral issues tend to be excluded from the discussion in an expert-based assessment.

> During the previous century a number of technologies –including window screens, private automobiles, sidewalk-free residential suburban streets and home air conditioning – contributed to the decline of face-to-face socializing and neighborliness in American residential communities. Now imagine a conventional, prospective, OTA-style study of one or more of these technologies, conducted at an appropriate date in the past. Let's suppose that the study is advised by a committee including – in addition to outside technical experts – representatives from organized stakeholder groups, such as leaders from a consumer organization, a labor union, an environmental group and several business trade organizations.

1.4 Limitations in Traditions Assessing Technologies

> The consumer representative would predictably focus on the potential cost, convenience and safety of these technologies. The worker representative would likely dwell especially on wages, job security and safety in the production process. The environmentalist might call attention to air pollution and the depletion of non-renewable resources. A representative of realtors might be concerned to prevent heavy-handed zoning or other regulations governing the development of suburban housing tracts. These are all reasonable concerns that merit inclusion in a TA study. But notice that no one on such a study advisory committee would be likely to shout, "Hey! What about the fact that all of these innovations could inhibit neighbors from talking and socializing with one another?" Absent any consideration of the possible effect of these technologies on social relations in daily life, there would presumably also be no attention to the follow-on question of how the technologically altered quality of community relations bears, in turn, on the basic ideals, structure and functioning of a democratic society. This is an example of how combining the views of even a very diverse range of organized stakeholder representatives, while helpful, can be insufficient to ensure that a TA study addresses the full range of significant social impacts and concerns. (Sclove 2010: 17; also see Sclove 1995: 3–9, 37–44 and 61–82)

This example shows how the public is unlikely to have a voice when only "organized groups" are engaged in the assessment of technologies. As Sclove points out, there are some values that are systematically neglected in transactions and negotiations among stakeholders with specific interests. Similarly, Swierstra and te Molder (2012) explain that some concerns about emerging technologies raised by citizens (for example, the question of "naturalness" in food industry) are discarded and minimized by technology developers. These concerns, typically non-quantifiable and ambiguous, are considered less important "soft" impacts that do not deserve attention. Currently, TA lacks the tools to include issues concerning the "greater good" in the debate or to discuss "soft" impacts like, for example, the impact of air-conditioning on neighborhood face-to-face social relations or the naturalness of food technologies.

Furthermore, as Science and Technology Studies have demonstrated, technological artifacts have a normative significance and moral connotation (Akrich 1992; Winner 1999; Latour and Venn 2002). They continuously challenge established societal norms, values and morality and require us to re-evaluate our normative frameworks (Keulartz et al. 2002). Take as an example, the well-documented case of the birth control pill (Keulartz et al. 2002). The pill's success is attributed to the self-discipline of the user who has to remember to consume it. While endowing women with more power in family planning, it also makes them more responsible (and blameworthy) in the case of unwanted pregnancy (Oudshoorn 2000). In re-distributing responsibilities and power relationships, artifacts also change our mentality and morals. For example, by separating sexuality from reproduction, the introduction of the pill contributes to a sexual revolution. It became easier for couples to engage in sexual activities outside wedlock, but it also promoted a birth planning mentality (Ketting 2000). The birth control pill forcefully illustrates how technologies and people's concepts, meanings and values mutually shape each other. These frameworks travel across practices and situations and are transported in several areas of moral deliberation: for example, the idea that a pregnancy must be planned is transported into the practice of abortion and plays an important role in people's decisions. Technology has the innovative and creative power to generate new moral challenges that require old convictions, concepts, meanings and values to be criticized and revised. Such co-shaping of technology and morality (Swierstra 2013) is often neglected in TA deliberative exercises.

Let us look at the case of CTA. The CTA approach is explicitly guided by three normative criteria: anticipation, earning and reflexivity (Schot and Rip 1997). Firstly, CTA assumes that it is better to *anticipate* possible paths of technological development, but emphasizes that such prospective thinking should be translated from the grammar of technological impact to the vocabulary of co-producing techno-social dynamics. Secondly, "*learning* must occur". Schot and Rip talk about "broad learning" to refer to the possible connections "between a range of aspects such as design options, user demands, and issues of political and social acceptability". Learning can also occur in the form of "deep learning". Deep learning can happen at two levels: a first-order learning that improves actors' capacity to work towards given goals and a second-order learning that provides a clarification of values. Finally,

> *reflexivity* is needed about the co-production of technology and society, to avoid to fall back into a naïve concept of impact, and about the role of the different actors in technological development (*ibidem*)

Anticipation, learning and reflexivity are presented as important conditions for improving the agency and deliberation of actors in a socio-technical world. These improvements are expected to lead to the ideal of a "better technology (in a better society)" (*ibid*: 256). However, there seems to be a tension between a deliberative ideal and the actual practice of CTA workshops. Whereas interacting workshops seem to operationalize the normative ideal of deliberative democracy by creating an idealized space of inquiry among different positions in a protected setting (Krabbenborg 2013), it is unclear how principles of democratic deliberation are taken up in these exercises. The spaces for interactions among stakeholders are described in CTA analyses as "negotiations" (Rip and Joly 2012) and CTA methods are in fact often presented as a way of "facilitating interfaces between the supply of science and technology and the demand for useful applications" (Merkerk and Smits 2008: 316). The "broadening" described by CTA analysts refers more to the enablement of stakeholders' capacity for effective decision making through increasing their understanding of socio-technical dynamics and creating spaces for collaborations and negotiations, rather than as a moral broadening. In the descriptions of CTA workshops there is no deliberation as an exploration of moral conflicts and evaluation of the moral acceptability (Grunwald 1999) of decisions or positions.

In principle, CTA aims at a "second-order deep learning" that requires stakeholders to clarify their firmly held values to each other. CTA declares that its methodology creates a space for different actors to challenge each other's positions, worldviews and values; however, in practice, it offers neither tools nor concepts to unravel and explicitly discuss normative conflicts among stakeholders. Descriptions of discussions in CTA workshops (cf. Robinson 2010) do not show an explicit exploration of the normative assumptions, values and norms when moral conflicts occur. Rather than focusing on values and norms, CTA workshops and analyses concentrate on alliances, linkages, de-alignments and the possible innovation paths that they can open up or close down. In this sense, it becomes clear how Grunwald's critique of the "normative deficit" in TA applies to CTA as well.

1.4.2 The Technological and Sociological Deficit in Institutional Ethics

Differently from TA parliamentary offices, (bio)ethical committees have a dedicated mandate of exploring the ethical implications of emerging technologies and evaluating their normative acceptability, however their analyses often lack sociological insight. Let us consider the case of the Opinions released by the European Group on Ethics (EGE) and in particular Opinion 20, which evaluates the emerging Information and Communication implants in the human body (European Group on Ethics 2005). Based on the philosophical concepts and principles of "autonomy", "personal integrity" and respect for human dignity, the EGE Opinion recommends policy makers adopt a precautionary approach towards implants used for surveillance or enhancement purposes and promote a broad social debate on these issues. Legislators are also called to deal with regulation concerning such implants. In line with scholarly approaches in applied ethics (Sandler 2013), these types of reports evaluate technologies through philosophical concepts and ethical normative theories. It has been questioned whether such a principalist based approach can offer good guidance for dealing with conflicts about values that emerge in social contexts in which different stakeholders hold divergent positions (Banta 2004; Reuzel 2004; Shelley Egan 2011). Furthermore, similarly to classical TA and HTA, this perspective does not acknowledge that societal values and framings and technology shape each other (Clausen and Yoshinaka 2004). In these reports promises of emerging technologies are taken at face value without a careful assessment of the material limitations and peculiarities that characterize them (Lucivero and Tamburrini 2007; Lucivero et al. 2011). This point has also been made with regard to ethical studies in the frame of ELSI programs.

The proliferation of ELSI/A funding programs has fostered the study of ethical aspects of emerging technologies that often offer only a general overview of ethical issues. In several cases, the contributions of ethicists (sometimes co-authoring with scientists) consist of reviews that attempt to compile inventories of ethical issues raised by technological applications (see for example, the case of nanotechnology Moor and Weckert 2004; Ebbesen and Jensen 2006; Lenk and Biller Andorno 2007). Although they have the merit of providing a broad overview of basic concerns surrounding emerging scientific fields and link new technologies to old issues whilst avoiding "reinventing the wheel", these reviews remain general and fail to have practical relevance for policy and technology development in reviewing bioethicists' discourse around pharmacogenetics, social scientist Adam Hedgecoe remarks that ethicists in ELSI projects engaged in "broad, but 'thin' reviews of potential ethical issues stay within the boundaries of ethical discourse set by academic and industry scientists" (Hedgecoe 2010: 176). In so doing, bioethicists fail to critically challenge scientific practice and discourse and step into the discursive field "mapped out" by technology developers (*ibidem*: 177) without criticizing the scientists' factual claims and "mirroring" the discourse of scientists promoting this technology.[27] Bioethicists in this way fall in an "intellectual capture" in the sense that they "begin to share the ideas, beliefs and goals of the group they are meant to be regulating" (*ibidem*: 178). Hedgecoe concludes that bioethicists thus allow a "smooth assimilation of pharmacogenetics in the clinical practice", rather than asking whether there *should* be such assimilation.

> This presupposition that technologies need to exist, if only in a promissory sense, before their ethical aspects can be debated, has led bioethicists to avoid discussion of issues beyond the pre-constructed discursive boundaries. (*ibid*: 179)

According to this strand of criticism, bioethicists uncritically accept scientists' expectations on the technology as an unquestioned premise of the ethical debate. This criticism does not apply to ethics per se. Ethicists' reflections on emerging technologies are often critical of emerging technologies. An example of this is provided by the work of political philosopher Michael Sandel.[28] Inspired by the discussions on genetic enhancements in 2004 he wrote a short essay for the *Atlantic Monthly* entitled "The Case against Perfection", extended into a short booklet in 2007. In this book, Sandel reflects on the promises in the field of genetics of treating and preventing debilitating diseases and its claims of enhancing human beings. Departing from the promises of genetics of treating and preventing diseases, the philosopher engages in a reflection on what is morally upsetting in the promises of genetic engineering for enhancement. He discards the usual arguments, instead appealing to the principles of autonomy, fairness and individual rights, because "this moral vocabulary is ill-equipped to address the hardest questions posed by genetic engineering" (Sandel 2004).[29] Moving away from this vocabulary he tries to look at genetic enhancement from a different perspective. He explains the moral stakes as, "[technologies for genetic enhancement] transform[ing] three key features of our moral landscape: humility, responsibility, and solidarity" (*ibidem*). He gives the example of pre-natal engineering. Promoting the desire of controlling offspring goes against the value of humility that is the basis of parents' unconditional

---

[27] In particular, Hedgecoe and Martin (2003) analyze the role of bioethical commentators in the co-construction of future scenarios around pharmacogenetics. They conclude that bioethics debate plays a role in creating visions, mobilizing resources and "anticipatory negotiation" over what is acceptable and what should be regulated. Bioethicists "through the anticipation of social and ethical problems and a critical engagement with the process of innovation, are also helping construct and shape the future" (*ibidem*: 357)

[28] Famous in Harvard for his course in political philosophy entitled "Justice", Sandel also co-teaches a course on "Ethics and Biotechnology" considering the ethical implications of several biotechnologies. Although he doesn't consider himself a professional ethicist, in 2001 he was invited to join the American President's Council of Bioethics.

[29] The full article is available at: http://www.theatlantic.com/past/docs/issues/2004/04/sandel.htm as an html version: page references are not available on this version

love. This excess of mastery fails to acknowledge the "giftedness of life" and the "openness to the unbidden" that are important in our society. Emerging genetic engineering is assessed against values that are important for our society, but whose importance is often undermined. Sandel's conclusions are that genetic engineering's "promise of mastery is flawed. It threatens to banish our appreciation of life as a gift, and to leave us with nothing to affirm or behold outside our own will" (*ibidem*).

Sandel discusses the desirability of the values and worldview promoted in techno-scientific promises. In questioning them he mobilizes the tradition of "virtue ethics" (or "good-life" ethics). In his ethical and conceptual reflection, Sandel points at issues that rarely appear in bioethical analysis. This work eludes the critique of "intellectual capture": he is not aligning with the enthusiastic hype of technology developers and instead raises a critical voice. However, Sandel's reflections are still taking the promises of scientists and engineers as a starting point. Even if it is able to go beyond the scientists' limited discourse and introduce some novel discussions, Sandel's analysis suffers the "operational deficit" of not creating space for consultation required by technology policy decision-making and does not take into account the constraints of the 'real world', i.e. how society, technology policy and technological development work (Grunwald 1999).

This type of ethical reflection falls into what Alfred Nordmann calls the "*if and then*" fallacy (Nordmann 2007) as the tendency of ethicists to begin their analysis by hypothesizing futuristic, visionary technologies and then to conclude with the hugely existential ethical impacts raised by these new technologies. In doing so, the hypothetical stance ("if") is usually downplayed and speculative scenarios are presented as imminent and pressing. The ethical debates related to brain-computer interfaces provide an illustration of this fallacy. These ethical arguments begin with postulating that such interfaces might become widespread in society, following which they turn immediately to discussing ethical concerns regarding the enhancement of human nature. In the process, this type of ethics discards the hypothetical status of the premise: "might" becomes "will" and the future is presented as on object of knowledge. While feeding unjustified hopes and fears, these discussions focus on an unknowable future and neglect present ethical issues: 'ethics leaps ahead of science' (Nordmann and Rip 2009). By drawing the public gaze towards unrealistic scenarios, ethicists unwittingly contribute to turning strategic promises about technologies into generally shared expectations. In this manner, the ethicist, rather than acting as a critical force, ends up supporting specific interest groups.

1.5 The Need for Integrating Ethical Inquiry in TA

The previous section has depicted the edges of a gap. On the one side there is TA "an analytic and democratic practice which aims to contribute to the timely formation of public and political opinion on societal aspects of science and technology" (van Est and Brom 2012) and, in its more participatory and interactive forms, strives for inclusion of all the relevant stakeholders in the deliberation of emerging technologies. Multifaceted TA approaches focus on the process of technology development and technology policy deliberation as well as on the production of policy reports. A "second generation" STS-oriented TA approach such as Constructive Technology Assessment strives for inclusion of all the relevant stakeholders in the deliberation of emerging technologies. In this approach, the broad/deep learning and reflexivity of actors in the innovation process is a major issue. In line with a (more or less explicit) ideal of deliberative democracy, different groups are involved in the discussions. This approach focuses on the process of technology development rather than its report-like product. The goal of offering evidence for decision-making to policy makers is accompanied (and sometimes replaced) by the goal of involving stakeholders in decisions on technology policies. CTA strives towards increasing stakeholders' awareness and understanding of the dynamics of the co-evolution of technology and society in the innovation process. I have discussed how some authors critically argue that TA approaches (including CTA) tend to put *empirical acceptance* before *normative acceptability*. According to this view, they do not create the space for an in-depth analysis of normative claims and assumptions and overlook a normative discussion of acceptability of the technology at stake. Furthermore, some issues about the desirability of emerging technologies are at risk of being systematically dismissed as "soft" by the stakeholders.

On the other side of the gap, ethical committees reports and some ELSI studies in bioethics and applied ethics aim at an explicit normative assessment of emerging technologies. They focus on technologies' desirability with respect to some moral principles or the way they endanger a vision of the good-life. In pointing out the potential conflicts with ethical principles fostered by future technologies, ethics approaches often seem to overlook the specificity of technical artifacts. Bioethicists' assessments risk being far removed from the actual state of the art in science and technology. Ethicists' distance from the actual process turns the debate into a speculative discussion about some innovations that might never come into existence. These studies are unable to feed the actual process of technology development because they fail to interact with actors in the "real" world (Shelley-Egan 2011) and to play a role in modulating morally relevant decisions during the innovation process.

This gap is depicted with a broad brush. Within the TA tradition some approaches have directly addressed the question of the implicit normativity and the importance of making it explicit. Within the broad ethics of technology tradition, applied ethics can be extremely context sensitive (as in the case of "empirical ethics" approaches in medical ethics and bioethics (Widdershoven et al. 2009; Willems and Pols 2010). There seems to be a general agreement that ethical inquiry should be integrated into TA in view of the productive learning that exists between these two traditions and the disciplines they foreground. As such, several authors have reflected on how integration could take place in practice (Roelofsen et al. 2008; Brey 2012):[30] ethical

---

[30] In 2001 a journal was created under the assumption that "technology assessment and ethics evaluation are to be developed into one methodologically integrated project" (Gethmann 2001):

Technology Assessment (Palm and Hansson 2006) and the ethical toolbox developed by the Ethical Bio-TA Tools project (Beekman and Brom 2007) specifically address the challenge of broadening TA to include moral issues, an exploration of stakeholders' meanings and visions, the unpacking of their core values and an analysis of their moral arguments.

In making the case for ethical technology assessment (eTA), Palm and Hansson (2006) propose a tool for identifying negative implications of emerging technologies at an early stage in continuous dialogue with technology development. This ethical technology assessment, integrated into the development of the new technology throughout its whole life-cycle, should be open to different perspectives and solutions, rather than proposing a unifying moral theory to evaluate emerging technologies. The authors identify a number of potential ethical issues associated with technologies that can be used as an "early warning system" or a "check-list" to indicate the need for an evaluation of the technology at an early stage.

1. Dissemination and use of information
2. Control, influence and power
3. Impact on social contact patterns
4. Privacy
5. Sustainability
6. Human reproduction
7. Gender, minorities and justice
8. International relations
9. Impact on human values.

(Palm and Hansson 2006: 555)

Though compelling and useful as a tool for early mapping of sensitive issues surrounding a certain technology development, this checklist approach falls short as it only focuses on the adverse effects of new technologies (Kiran et al. 2015).[31] Furthermore, the list pre-establishes the set of relevant ethical issues, not accounting for the fact that novel technologies disrupt current moral orders (Keulartz et al. 2002; Grunwald 2004) and raise new ethical controversies. In offering a fixed set of "usual suspects" this approach does not account for the fact that morality and technology co-evolve mutually shaping one another (Swierstra et al. 2009; Swierstra 2013).

This aspect of the co-shaping of technology and society has been frequently brought up in the discussion regarding the integration of ethics in TA. Starting from the premise that the integration of ethical inquiry into TA is not an easy enterprise, several authors have made explicit the moral aspects of technologies and drawn attention to the increasing awareness that developing, implementing and using a technology is not value free (see for example, Oortwijn et al. 2004). The social context – including the use, the policy making environment, participatory approaches

---

Poiesis and Praxis, the International journal on the Ethics of Science and Technology Assessment. An entire issue in 2004 addressed the need of integrating ethics in (H)TA.

[31] In this recent contribution, Kiran and colleagues (2015) call for the need to go beyond checklist approaches as a way of including ethical reflection in (C)TA, proposing an approach which is in line with the one described in this book.

and conflicts among social groups – sets the norms and values that influence and are influenced by the technology under investigation (Clausen and Yoshinaka 2004). The co-shaping of technology and social norms appears to have a central role in the evaluation of the moral aspects of a technology, as it is recognized that technologies are intrinsically normative. This shifts the focus from the future impacts of a technology, to the way it is shaped in the present and furthermore points out that there are no pre-given ethical questions, as their framing is itself the product of social interaction (Reuzel et al. 2004). The goal of ethical inquiry is therefore to open issues of scrutiny, debate and redefinition by exploring the ethical issues relevant to stakeholders and the need to shift the socio-technical agenda.[32] Similarly, the project Ethical Bio-TA aims at identifying "practical instruments that can be used (tools) in order to support debates and deliberative structures for a systematic engagement with ethical issues" (Beekman and Brom 2007: p4). The integration of ethics in TA is again not seen as a way of privileging a specific normative approach or philosophical tradition, but as a facilitation of discussions on values in a pluralistic society in which different participants have different ethical perspectives. Although these approaches offer very compelling arguments, tools and observations, it is often unclear what the normative framework that would justify such approaches is (although a few single contributions offer some guidance in this respect, see for example Grunwald 2004).

Holding a similar perspective, Keulartz and colleagues programmatically propose a normative framework to investigate the ethical aspects in a technological culture, which is based on pragmatist philosophy (2002, 2004). Positioning pragmatist ethics as a normative approach that stands in the middle ground between a technologically blind applied ethics approach and a normatively lacking STS methodology, the authors seek to acknowledge pluralism in deliberations around technologies while avoiding relativism. Pragmatist ethics theorizes the possibility of dynamic changes in morality and hold a deliberative democratic ideal according to which decision-making is achieved through public participation and consultation and questioning of the established order via social inquiry (Dewey 1981; Habermas 1990). Such an approach offers a normative framework that not only acknowledges societal value to conflicts among stakeholders, but also highlights the need to question the very foundations of established values and moral frameworks within a specific normative context (or societal practice).

According to Dewey, morality[33] is a dynamic process characterized by continuous adjustments. Legitimized by tradition and historical continuity, some values,

---

[32] In line with this approach is the "Interactive" TA approach (Grin et al. 1997), which focuses on the interaction between a variety of stakeholders holding different "frames of meaning" (Grin and van de Graaf 1996) and learning from each other's. The "Vision Assessment" approach (Grin and Grunwald 2000) more directly aims at creating spaces for deliberation by articulating the normative content of technological visions. Because of its focus on future oriented visions and projections, this approach will be discussed in the next chapter.

[33] Interestingly, the word 'morality' comes from the Latin *mos-moris*, custom. In ancient Rome, the *mos maiorum* (in English, ancestral custom or fathers' custom) was an unwritten code of habits, principles, behavioral models and practices that affected the private, political and military life in

principles and customs are taken for granted in a certain society. However, they are continuously evolving because of cultural and social interactions that demand a continuous interpretation of the values and their role in society. *Ethical* discussions on the legitimacy of some values, their meaning and their priority, trigger adjustments in morality. The morality shared by a community is thus expressed in the routines and habits of this community that are not questioned unless there is a problem. For example, when a decision is hard to make or a conflict occurs among different values. In cases like these, a community becomes more aware of the "background of convictions", experienced as unproblematic up to that point. Values are discussed, new vocabularies and concepts are created, a "workable solution" is constructed, and a new order is established (Keulartz et al. 2002: 13). Furthermore, moral elements like principles, values or norms are often transposed beyond their original context of invention into other areas where similar problems play a role.[34] In proposing the appropriateness of pragmatist ethics as a framework for ethical inquiry in a technological context, Keulartz et al. reconcile the context sensitivity and moral anti-foundationalism of classical pragmatist ethics with Science and Technology Studies' work on the co-shaping mechanisms between technology and society. As new devices intimately interact with human beings and challenge existing established traditions and as values co-evolve with technologies and have to be reevaluated; ethical controversies are likely to occur (Swierstra 2015).[35]

That technologies shape many aspects of human experiences and societal context as much as they are shaped by societal values and framings, is a fact acknowledged in the recent textbook "Ethics and Emerging Technologies" (Sandler 2013). Collecting several perspectives and issues related to technologies, the textbook offers a framework for the ethical analysis of emerging technologies that requires "identifying any benefits" and "extrinsic concerns" as well as

---

Rome. The virtues and values prescribed by this code were accepted by the community because of the authority of the tradition. Furthermore, the written social norms were derived from this code. Despite coming from uncertain origin, the traditional values in the *mos maiorum* were considered as a stable basis for Roman identity. Some authors like Cato vehemently argued that observing the fathers' custom was essential to the wellbeing of the Roman Republic. However, in later years of the Republic, the lower (plebeian) social class undermined the conservative principle of the *mos* in order to achieve social and political reforms. Furthermore, when the Roman Empire had to manage a large conglomerate of different populations with very different customs, practices and values, the *mos regionis* (regional custom) was set alongside the *mos maiorum*. Beside this variability in the social and geographical dimensions, the traditional custom was also challenged in the temporal dimension.

[34] An example of the transposition of concepts beyond their original context is provided by (Swierstra et al. 2009). The meaning of the principle of "autonomy" has changed and has been transported to different problem areas: "This principle was first coined to elucidate the precarious political status of the fifteenth century Italian city-state, played an important role in religious controversies, resurfaced in Rousseau's political philosophy, was elevated by Kant to take central stage in morality, and has in the last decades finally reached public prominence in the field of medical ethics" (132)

[35] The normative purport of pragmatist ethics and its contribution to addressing the normative deficit in TA will be further discussed in Chap. 7.

"intrinsic concerns that the technology is likely to raise" in addition to conducting "power" and "form of life" analyses (pp. 19–20). The authors do not examine the details of this framework but point out there is always a speculative element to these analyses:

> When technologies are not yet fully developed we will not know precisely what their features are or how they work; and when they are not widely disseminated we will not know precisely what their impacts are (p 20).

The authors highlight the importance of "informed, well measured anticipation" as opposed to "wild speculation". However, the questions remain of how to assure informed, well measured anticipation? How to identify benefits and concerns in a non-wildly speculative way? And what for?

These aspects have been addressed in recent debates on the legitimacy of the so-called "ELSA" approaches to the study of nanotechnology. According to Nordmann and Rip (2009), in order to bridge the gap between futuristic ethical speculation and actual technology development, ethicists should precede an ethical evaluation of (nano)technology with "reality checks". Such "reality checks" consist of assessing the quality of scientific expectations in order to establish whether they refer to feasible technical developments worthy of ethical reflection. This would close the gap between ethical scenarios and current science. As Nordmann explains in another article, "nanotechnological and other techno-scientific prospects suffer from the failure to distinguish physical possibility (all that does not contradict outright the laws of nature) and technical possibility (all that humans can build)" (Nordmann 2007b: 4). By assessing the quality of expectations, the ethicist would critically appraise the conditions of technical possibility rather than taking them for granted.

Nordmann and Rip's commentary elicited several responses about the legitimacy and desirability of "speculation" in ethical reflections on emerging technologies (Lucivero et al. 2011; Michelfelder 2011; Ferrari and Grunwald 2011). Grunwald (2010) argues that it would be a mistake to cluster all the reflections on the ethics of emerging science and technology under the same label. "Nano/gene/robo/computer/bio/neuro-ethics" are umbrella terms that include different types of studies. Under this label, there are studies that belong to "applied ethics" which deal with concrete questions to be considered in the context of regulating emerging technologies. For example whether nanoparticles should be included in food or, what are the equity issues in benefitting from nanotechnologies. These studies come later in the technology development process, when all the relevant decisions have already been made. Therefore they have less of a role in shaping technological development. However, other types of studies are also included in this generic "ethics" label, such as philosophical work on human enhancement or human-technology relationships. These studies have joined the STS move towards upstream engagement and have different purposes and methodologies from applied ethics. Rather than addressing concrete and existing ethical problems, this "explorative nano-philosophy" contains philosophical reflections of different types: epistemological, anthropological and hermeneutical. Whereas applied ethics aims at directing policy makers or other social actors towards some actions (for example, regulating a technological field) and must, for this reason, be concerned with concrete problems of the "here and now", *explorative* nano-philosophy has a more diverse set of purposes. For example,

> early thoughts about synthetic biology or human enhancement serve rather to promote the conceptional understanding and clarification of the issues from a normative perspective or to facilitate the development of clear terminology and ethical alternatives, but without there being anything immediate to be regulated. (*Ibid*: 96)

Explorative philosophy can prepare science and society for issues that could emerge in the future, draw attention to conceptual or normative gaps in today's visions of tomorrow, and provide interpretative tools to "learn something 'about and for us' today" (*ibid*: 97). The purpose of an ethical/philosophical reflection on emerging technologies is not only to account for contextual and existing ethical dilemmas, but also to engage in more "explorative" reflections. As Grunwald points out, explorative philosophy is in itself an emerging field and as such needs a methodological and epistemological foundation.

1.6 Between Grounding and Exploring: The Contribution of This Study

In Plato's myth, Theuth presents his inventions explaining why they are desirable; however, King Thamus claims a legitimate authority to assess the desirability of these inventions, because the inventors are not the best judges of their brainchildren. What the best practice is for assessing emerging technologies varies across times, societies and disciplines. It can be argued that two trends have been dominant in modern Western societies: a democratization of the process of assessing technologies and an inclusion of "broader" ethical concerns in such assessments. As some commentators point out, an assessment of the desirability of emerging technologies has a normative significance that has to be explicitly addressed. Technology Assessment activities and theories are sometimes lacking in this respect. The fields of applied ethics and bioethics do address normative questions in the evaluation of emerging technologies, but they are often insufficiently critical of the empirical grounds of their analysis.

In addressing the question of how to integrate ethical inquiry in TA – i.e. how to account for societal contexts, techno-social co-shaping and stakeholders' conflicts while exploring the moral dimensions inherent in new technologies or actors' controversies – issues of emergence, uncertainty and dynamic evolution deserve attention. As we will see in the next chapter, the object of the assessment consists of technologies that are still emerging and exist only as "expectations". This point has been foregrounded within the debate on speculative ethics. As explained above, this debate unfolds in opposing directions. On the one hand, critics indicate a risk of questioning the desirability of emerging technologies that will never be there. Such a speculative ethical reflection would reinforce the hype surrounding the development of a technology and could be instrumentally used to raise fear or enthusiasm in the public domain. On the other hand, it is emphasized that a reflection that goes beyond a here-and-now discourse is important. Such reflection should be able to explore scenarios of the future with the goal of assessing the normative acceptability of present choices and decisions. Whereas the critique of speculative ethics points out the need for "reality checks" in order to draw attention to the "here and now" issues, the "explorative (nano)philosophy" argument indicates that speculative reflections on emerging technologies can still be valuable. How do we engage in an epistemological analysis of the foundations of such an enterprise and assess the quality of the information about the technology in question?

To further understand this point, let's go back to the Platonic myth we started with. Thamus hears about the alphabet directly from its inventor; how can Thamus understand how the alphabet works? How will Egyptians use this "invention'? How can Thamus anticipate what it will do once it is embedded in Egyptian society? At an early stage of a technology's development, many variables and uncertainties make it difficult to anticipate the final technology as a product, let alone to evaluate its desirability. The fundamental question here is: how to deal with these uncertainties while technologies are still emerging?

Technology assessment is an attempt of modern societies to manage uncertainties and contribute to strategic governance of the process of innovation. In this context, a future that has not yet been realized turns into an object of analysis and modulation. This upstream assessment concerns expectations of a future technology, not yet realized. As the next chapter will explain, these expectations are visions of a future state of affairs in which a certain artifact is integrated into social practices and interacts with human beings in a certain way. An epistemological analysis is needed in order to analyze the possibilities for an assessment of emerging technologies that aims at engaging in normative discussions about their moral value. Building on studies that have discussed expectations and visions of emerging technologies as well as the need to gather anticipatory knowledge and make decisions on uncertain futures,[36] the next chapters will discuss how the moral assumptions and normative conflicts in stakeholders' and artifacts' visions can be brought forward for discussion.

As the discussion on speculative ethics has made clear, it is important to establish "criteria and procedures for better being able to distinguish between mere speculations and more plausible futures" (Grunwald 2010: 95). This appraisal of the epistemological foundations of expectations, while assessing their viability, should also transcend the logic of "here and now" and explore novel and distinctive issues raised by the expectations of new technologies. This study aims at showing that the requirements of grounding hype and avoiding mere speculation and an explorative reflection on the future are not mutually exclusive. On the contrary, they are both necessary criteria that should guide the upstream discourse on the ethical and social aspects of emerging technologies. This book presents a methodological approach

---

[36] As for example Constructive Technology Assessment and Vision Assessment, but also Foresight Studies.

for analyzing and articulating the moral reasons, meanings and commitments implied in expectations concerning emerging technologies. I claim that such an approach is a pre-condition for *any* ethical assessment of emerging technologies.

References

Abels, Gabriele. 2007. Citizen involvement in public policy-making: Does it improve democratic legitimacy and accountability? The case of pTA. *Interdisciplinary Information Sciences* 13(1): 103–116.

Akrich, M., and B. Latour. 1992. A summary of a convenient vocabulary for the semiotics of human and nonhuman assemblies. In *Shaping technology, building society: Studies in sociotechnical change*, ed. W. Bijker and J. Law, 259–264. Cambridge, MA: MIT Press.

Banta, H. David (Coordinator). 1997. Report from the EUR-ASSESS project. *International Journal of Technology Assessment Health Care* 13: 131–340.

Banta, H. David. 2004. Foreword. *Poiesis & Praxis: International Journal of Technology Assessment and Ethics of Science*, 2(2–3):93–95.

Banta, H. David, and Seymour Perry. 1997. A history of ISTAHC: A personal perspective on its first 10 years. *International Journal of Technology Assessment in Health Care* 13(03): 430. Cambridge University Press.

Beauchamp, Tom L. 1995. Principlism and its alleged competitors. *Kennedy Institute of Ethics Journal* 5(3): 181–198. doi:10.1353/ken.0.0111. The Johns Hopkins University Press.

Beauchamp, T.L., and J.F. Childress. 2009. *Principles of biomedical ethics* [1973] ed. New York: Oxford University Press.

Beekman, Volkert, and Frans W.A. Brom. 2007. Ethical tools to support systematic public deliberations about the ethical aspects of agricultural biotechnologies. *Journal of Agricultural and Environmental Ethics* 20(1): 3–12.

Bijker, W.E. 1995. *Of bicycles, bakelites, and bulbs: Toward a theory of sociotechnical change*. Cambridge, MA: MIT Press.

Blok, Anders. 2007. Experts on public trial: On democratizing expertise through a Danish consensus conference. *Public Understanding of Science* 16(2): 163–182.

Brey, P. 2012. Anticipatory ethics for emerging technologies. *NanoEthics* 6(1): 1–13.

Broerse, J.E.W., and J.F.G. Bunders. 2000. Requirements for biotechnology development: The necessity for an interactive and participatory innovation process. *International Journal of Biotechnology* 2(4): 275–296.

Bush, Vannevar. 1945. *Science, the endless frontier: A report to the president*. Washington, DC: Government Printing Office.

Callon, Michel. 1986. Some elements of a sociology of translation: Domestication of the Scallops and the Fishermen of St. Brieuc Bay. In *Power, action, and belief: A new sociology of knowledge*, vol. 32, ed. John Law, 196–223. London: Routledge.

Callon, M., J. Law, and A. Rip (eds.). 1986. *Mapping the dynamics of science and technology: Sociology of science in the real world*. London: Macmillan.

Calvert, Jane, and Paul Martin. 2009. The role of social scientists in synthetic biology. Science & society series on convergence research. *EMBO Reports* 10(3): 201–204. EMBO Press.

Chambers, Simone. 2003. Deliberative democratic theory. *Annual Review of Political Science* 6(1): 307–326.

Clausen, Christian und Yutaka Yoshinaka. 2004. Social shaping of technology in TA and HTA. *Poiesis & Praxis: International Journal of Technology Assessment and Ethics of Science* 2(2–3): 221–246.

Clouser, K. Danner. 1995. Common morality as an alternative to principlism. *Kennedy Institute of Ethics Journal* 5(3): 219–236. The Johns Hopkins University Press.

Collingridge, D. 1980. *The social control of technology*. Milton Keynes: Open University Press.
Decker, Michael, and Miltos Ladikas. 2004. *Bridges between science, society and policy: Technology assessment – methods and impacts*. Berlin: Springer Science & Business Media.
Dewey, John, and Jo. Ann Boydston. 1981. *The later works, 1925–1953*. Carbondale/London: Southern Illinois University Press/Feffer & Simons.
Dryzek, John S. 2000. *Deliberative democracy and beyond liberals, critics, contestations*. New York: Oxford University Press.
Ebbesen, M., and T.G. Jensen. 2006. Nanomedicine: Techniques, potentials, and ethical implications. *Journal of Biomedicine and Biotechnology* 2006: 1–11.
European Group on Ethics. 2005. Ethical aspects of ICT implants in the human body. Opinion 20. Available at http://ec.europa.eu/european_group_ethics/docs/avis20_en.pdf.
European Group on Ethics. 2007. Ethical aspects of nanomedicine. Opinion 21.
Ferrari, A., and A. Grunwald. 2011. Visions and ethics in current discourses on human enhancement, third annual conference of the society for the study of nanoscience and emerging technologies (S.NET), Tempe, Arizona/USA, 07–10 Nov 2011.
Fischer, F., and J. Forester. 1993. *The argumentative turn in policy analysis and planning*. Durham: Duke University Press.
Fisher, Erik, Roop L. Mahajan, and Carl Mitcham. 2006. Midstream modulation of technology: Governance from within. *Bulletin of Science Technology Society* 26(6): 485–496.
Funtowicz, Silvio O., and Jerome R. Ravetz. 1993. The emergence of post-normal science. In *Science, politics and morality SE – 6*, ed. Schomberg René Von, 17:85–123. Theory and Decision Library. Dordrecht: Springer.
Gethmann, Carl Friedrich. 2001. A new journal is launched. *Poiesis & Praxis* 1(1): 1–2.
Gibbons, Michael, Camille Limoges, Helga Nowotny, Simon Schwartzman, Peter Scott, and Martin Trow. 1994. *The new production of knowledge: The dynamics of science and research in contemporary societies*. London: Sage.
Grin, J., and H. van de Graaf. 1996. Technology assessment as learning. *Science, Technology & Human Values* 21(1): 72–99.
Grin, J., and A. Grunwald. 2000. *Vision assessment: Shaping technology in 21st century society*. New York: Springer.
Grin, John, H. van de Graaf, and Rob Hoppe. 1997. *Technology assessment through interaction. A guide*. The Hague: Rathenau Institute. Available at https://www.researchgate.net/publication/254753158_Technology_Assessment_through_interaction._A_guide
Grunwald, A. 1999. Technology assessment or ethics of technology? *Ethical Perspectives* 6(2): 170.
Grunwald, Armin. 2004. The normative basis of (health) technology assessment and the role of ethical expertise. *Poiesis & Praxis: International Journal of Technology Assessment and Ethics of Science* 2(2–3): 175–193.
Grunwald, Armin. 2009. Technology assessment: Concepts and methods. In *Philosophy of technology and engineering sciences*, ed. Anthonie W.M. Meijers. North Holland: Elsevier.
Grunwald, Armin. 2010. From speculative nanoethics to explorative philosophy of nanotechnology. *NanoEthics* 4(2): 91–101. Springer, Netherlands.
Grunwald, Armin. 2011. Responsible innovation: Bringing together technology assessment, applied ethics, and STS research. November. IET. http://run.unl.pt/handle/10362/7944.
Guston, D.H., and D. Sarewitz. 2002. Real-time technology assessment. *Technology in Society* 24(1–2): 93–109.
Gutmann, Amy, and Dennis Thompson. 2002. Deliberative democracy beyond process. *Journal of Political Philosophy* 10(2): 153–174.
Habermas, J. 1990. *Moral consciousness and communicative action*. Cambridge, MA: MIT Press.
Hedgecoe, A. 2010. Bioethics and the reinforcement of socio-technical expectations. *Social Studies of Science* 40(2): 163–186.
Hedgecoe, A., and P. Martin. 2003. The drugs don't work: Expectations and the shaping of pharmacogenetics. *Social Studies of Science* 33(3): 327–364.

Hennen, Leonhard. 2012. Why do we still need participatory technology assessment? *Poiesis & Praxis: International Journal of Ethics of Science and Technology Assessment* 9(1–2): 27–41.
Hoeyer, Klaus. 2006. "Ethics wars": Reflections on the antagonism between bioethicists and social science observers of biomedicine. *Human Studies* 29(2): 203–227.
Hofmann, Bjørn. 2005. On value-judgements and ethics in health technology assessment. *Poiesis & Praxis* 3(4): 277–295.
Hofmann, Bjørn Morten. 2008. Why ethics should be part of health technology assessment. *International Journal of Technology Assessment in Health Care* 24(4): 423–429.
Jasanoff, Sheila. 2004. *States of knowledge: The co-production of science and the social order*. London: Routledge.
Jensen, Casper Bruun. 2005. Citizen projects and consensus-building at the Danish board of technology. *Acta Sociologica* 48(2): 221–235.
Jonsen, Albert R. 1986. Casuistry and clinical ethics. *Theoretical Medicine* 7(1): 65–74.
Joss, Simon, and Sergio Bellucci. 2002. *Participatory technology assessment: European perspectives*. London: Centre for the Study of Democracy.
Kearnes, Matthew und Brian Wynne. (2007). On nanotechnology and ambivalence: The politics of enthusiasm. *NanoEthics* 1(2): 131–142.
Ketting, E. 2000. De Invloed van Orale Anticonceptie Op de Maatschappij. *Nederlands Tijdschrift Voor Geneeskunde* 144(6): 283–286.
Keulartz, J., M. Schermer, M. Korthals, and T. Swierstra (eds.). 2002. *Pragmatist ethics for a technological culture*. Deventer: Kluwer.
Keulartz, Jozef, Maartje Schermer, Michiel Korthals, and Tsjalling Swierstra. 2004. Ethics in technological culture: A programmatic proposal for a pragmatist approach. *Science, Technology & Human Values* 29(1): 3–29.
Kiran, Asle H., Nelly Oudshoorn, and Peter-Paul Verbeek. 2015. Beyond checklists: Toward an ethical-constructive technology assessment. *Journal of Responsible Innovation*. doi:10.1080/23299460.2014.992769, January. Routledge.
Klüver, Lars, Michael Nentwich, Walter Peissl, Hele Torgersen, Fritz Gloede, Leonhard Hennen, Josée van Eijndhoven, Rinie van Est, Simon Joss und S Belluci. 2000. European participatory technology assessment. *Participatory methods in technology assessment and technology decision-making*. Copenhagen: The Danish Board of Technology.
Krabbenborg, Lotte. 2013. *Involvement of civil society actors in nanotechnology: Creating productive spaces for interaction*. Groningen: University of Groningen.
Kuhse, H., and P. Singer. 2012. *A companion to bioethics*. Oxford/Malden: Blackwell.
Latour, B. 1987. *Science in action: How to follow scientists and engineers through society*. Cambridge, MA: Harvard University Press.
Latour, Bruno. 2005. Reassembling the social-an introduction to actor-network-theory. In *Reassembling the social-an introduction to actor-network-theory*, ed. Latour Bruno, 316. *Foreword by Bruno Latour*. Oxford: Oxford University Press.
Latour, Bruno, and Couze Venn. 2002. Morality and technology: The end of the means. *Theory, Culture & Society* 19(5–6): 247–260.
Law, John. 1986. On the methods of long-distance control: Vessels, navigation and the Portuguese route to India. In *Power, action and belief: A new sociology of knowledge*, 234–263. London: Routledge.
Lehoux, P. 2006. *The problem of health technology: Policy implications for modern health care systems*. London: Routledge.
Lehoux, Pascale, and Bryn Williams-Jones. 2007. Mapping the integration of social and ethical issues in health technology assessment. *International Journal of Technology Assessment in Health Care* 23(1): 9–16.
Lenk, C., and N. Biller Andorno. 2007. Nanomedicine-emerging or re-emerging ethical issues? A discussion of four ethical themes. *Medicine, Health Care & Philosophy* 10(2): 173–184.
Lucivero, Federica, and Guglielmo Tamburrini. 2007. Ethical monitoring of brain-machine interfaces. *Ai & Society* 22(3): 449–460.

Lucivero, Federica, Tsjalling Swierstra, and Marianne Boenink. 2011. Assessing expectations: Towards a toolbox for an ethics of emerging technologies. *NanoEthics* 5(2): 129–141.
Mepham, B. 2001. Novel foods. In *The concise encyclopedia of the ethics of new technologies*, ed. R.F. Chadwick. San Diego: Academic.
Merton, R.K. 1973. The normative structure of science (1942). In *The sociology of science. Theoretical and empirical investigations*, 267–278. Chicago: University of Chicago Press.
Michelfelder, Diane. 2011. Dirty hands, speculative minds, and smart machines. *Philosophy & Technology* 24(1): 55–68.
Moor, J., and J. Weckert. 2004. Nanoethics: Assessing the nanoscale from an ethical point of view. In *Discovering the nanoscale*, ed. D. Baird, A. Nordmann, and J. Schummer. Amsterdam/Washington, DC: Ios Press.
Nordmann, A. 2007. If and then: A critique of speculative NanoEthics. *NanoEthics* 1(1): 31–46.
Nordmann, A., and A. Rip. 2009. Mind the gap revisited. *Nature Nanotechnology* 4(5): 273–274. Nature.
Nowotny, Helga, Peter Scott, and Michael T. Gibbons. 2001. *Re-thinking science: Knowledge and the public in an age of uncertainty*. London: Wiley.
Oortwijn, Wija, Rob Reuzel, and Michael Decker. 2004. Introduction. *Poiesis & Praxis: International Journal of Technology Assessment and Ethics of Science* 2(2–3): 97–101.
Oudshoorn, N. 2000. The co-construction of contraceptive technologies and users. In *Bodies of technology. Women's involvement in reproductive medicine*, ed. A. Saetnan, N. Oudshoorn, and M. Kirejczyk. Ohio: Ohio University Press.
Oudshoorn, N., and T.J. Pinch. 2003. *How users matter: The co-construction of users and technologies*. Cambridge, MA: MIT Press.
Owen, R., P. Macnaghten, and J. Stilgoe. 2012. Responsible research and innovation: From science in society to science for society, with society. *Science and Public Policy* 39(6): 751–760.
Palm, E., and S.O. Hansson. 2006. The case for ethical technology assessment (eTA). *Technological Forecasting & Social Change* 73(5): 543–558.
Pellegrino, Edmund D. 1995. Toward a virtue-based normative ethics for the health professions. *Kennedy Institute of Ethics Journal* 5(3): 253–277. The Johns Hopkins University Press.
Pinch, Trevor J., and Wiebe E. Bijker. 1984. The social construction of facts and artefacts: Or how the sociology of science and the sociology of technology might benefit each other. *Social Studies of Science* 14(3): 399–441.
Plato. 1973. The Phaedrus of Plato. Transl. and ed. W.H. Thompson. New York: Arno Press.
Reuzel, Rob. 2004. Interactive technology assessment of paediatric cochlear implantation. *Poiesis & Praxis* 2(2–3): 119–137.
Reuzel, Rob, Wija Oortwijn, Michael Decker, Christian Clausen, Pedro Gallo, John Grin, Armin Grunwald, Leo Hennen, Gert Jan van der Wilt, and Yutaka Yoshinaka. 2004. Ethics and HTA: Some lessons and challenges for the future. *Poiesis & Praxis* 2(2–3): 247–256.
Rip, Arie. 2009. Technology as prospective ontology. *Synthese* 168(3): 405–422.
Rip, Arie, and Haico te Kulve. 2008. Constructive technology assessment and socio-technical scenarios. *Nanotechnology* 1: 49–70.
Rip, Arie, and Pierre-Benoit Joly. 2012. Emerging spaces and governance. *A position paper for Eu-SPRI forum*. Available at https://www.researchgate.net/profile/Andy_Stirling/publication/263962630_Emerging_Spaces_and_Governance_A_position_paper_for_EU-SPRI/links/00b4953c67201913bd000000.pdf
Rip, A., T.J. Misa, and J. Schot. 1995. *Managing technology in society: The approach of constructive technology assessment*. London/New York: Pinter.
Robinson, D.K.R. 2009. Co-evolutionary scenarios: An application to prospecting futures of the responsible development of nanotechnology. *Technological Forecasting and Social Change* 76(9): 1222–1239.
Robinson, D.K.R. 2010. *Constructive technology assessment of emerging nanotechnologies experiments in interactions*. Enschede: Proefschrift Universiteit Twente.

References

Roelofsen, A., et al. 2008. Exploring the future of ecological genomics: Integrating CTA with vision assessment. *Technological Forecasting and Social Change* 75(3): 334–355.

Sandel, M. 2004. The case against perfection, *Atlantic Monthly* 293(3): 51–62. Available at http://www.theatlantic.com/past/docs/issues/2004/04/sandel.htm

Sandler, Ronald. 2013. *Ethics and emerging technologies*. New York: Palgrave Macmillan.

Schot, J., and A. Rip. 1997. The past and future of constructive technology assessment. *Technological Forecasting and Social Change* 54(2–3): 251–268.

Sclove, R. 1995. *Democracy and technology*. New York: Guilford Press.

Sclove, R. 2010. *Reinventing technology assessment: A 21st century model*. Washington, DC: Science and Technology Innovation Program.

Shelley Egan, C. 2011. *Ethics in practice: Responding to an evolving problematic situation of nanotechnology in society*. Enschede: Proefschrift Universiteit Twente.

Smits, R.E.H.M., and A.J.M. Leyten. 1991. *Technology assessment: Waakhond of speurhond? Naar een integraal technologiebeleid*. Zeist: Kerckebosch.

Smits, Ruud, Jos Leyten, Pim Den Hertog, und Pim Hertog. 1995. Technology assessment and technology policy in Europe: New concepts, new goals, new infrastructures. *Policy Sciences* 28(3): 271–299.

Smits, Ruud, Rutger van Merkerk, David H. Guston, Daniel Sarewitz, and R. van Merkerk. 2008. The role of TA in systemic innovation policy. Intelligence. Innovation Studies Utrecht (ISU) Working paper series.

Stegmaier, Peter. 2009. The rock 'N' roll of knowledge co-production. Science & society series on convergence research. *EMBO Reports* 10(2): 114–119.

Swierstra, T. 1997. From critique to responsibility. *Society for Philosophy and Technology* 3(1): 68–74.

Swierstra, Tsjalling. 2013. Nanotechnology and technomoral change. *Etica e Politica* 15: 200–219.

Swierstra, Tsjalling. 2015. Identifying the normative challenges posed by technology's 'soft' impacts. *Etikk I Praksis – Nordic Journal of Applied Ethics* 9(1): 5–20.

Swierstra, Tsjalling, and Arie Rip. 2007. Nano-ethics as NEST-ethics: Patterns of moral argumentation about new and emerging science and technology. *NanoEthics* 1(1): 3–20.

Swierstra, T., D. Stemerding, and M. Boenink. 2009. Exploring techno-moral change: The case of the obesity pill. In *Evaluating new technologies*, The international library of ethics, law and technology, vol. 3, ed. Sollie Paul and Düwell Marcus, 119–138. Dordrecht: Springer.

Swierstra, T., and H. te Molder. 2012. Risk and soft impacts. In *Handbook of risk theory*, ed. S. Roeser, R. Hillerbrand, M. Peterson, and P. Sandin, 1050–1066. Dordrecht: Springer.

te Kulve, H. 2011. *Anticipatory interventions and the co-evolution of nanotechnology and society*. Enschede: Proefschrift Universiteit Twente.

Ten Have, Henk. 2004. Ethical perspectives on health technology assessment. *International Journal of Technology Assessment in Health Care* 20(01): 71–76. Cambridge University Press.

Tran, Thien A., and Tugrul Daim. 2008. A taxonomic review of methods and tools applied in technology assessment. *Technological Forecasting and Social Change* 75(9): 1396–1405.

Van Eijndhoven, J.C.M. 1997. Technology assessment: Product or process? *Technological Forecasting and Social Change* 54(2–3): 269–286.

van Est, R., and Brom, F. 2012. Technology assessment as an analytic and democratic practice. *Encyclopedia of Applied Ethics,* 2e Chapter 10 on "Technology assessment".

Van Merkerk, R.O. 2007. *Intervening in emerging nanotechnologies: A CTA of lab on a chip technology*. Utrecht: Utrecht University, Royal Dutch Geographical Society.

van Merkerk, R., and D. Robinson. 2006. Characterizing the emergence of a technological field: Expectations, agendas and networks in lab-on-a-chip technologies. *Technology Analysis & Strategic Management* 18(3/4): 411–428.

van Merkerk, R., and R. Smits. 2008. Tailoring CTA for emerging technologies. *Technological Forecasting and Social Change* 75(3): 312–333.

Van Merkerk, Rutger O., and Harro van Lente. 2005. Tracing emerging irreversibilities in emerging technologies: The case of nanotubes. *Technological Forecasting and Social Change* 72(9): 1094–1111.

Van Oudheusden, Michiel. 2014. Where are the politics in responsible innovation? European governance, technology assessments, and beyond. *Journal of Responsible Innovation* 1(1): 67–86.

von Hippel, E. 1988. *The sources of innovation*. New York: Oxford University Press.

Widdershoven, Guy, Tineke Abma, and Bert Molewijk. 2009. Empirical ethics as dialogical practice. *Bioethics* 23(4): 236–248.

Willems, Dick, and Jeannette Pols. 2010. Goodness! The empirical turn in health care ethics. *Medische Antropologie* 1(22): 161–170.

Winner, Langdon. 1999. Do artifacts have politics? In *The social shaping of technology*, vol. 29, ed. Donald MacKenzie and Wajcman Judy, 28–40. Buckingham: Open University Press.

Zwart, Hub, Laurens Landeweerd, and Arjan van Rooij. 2014. Adapt or perish? Assessing the recent shift in the European research funding arena from 'ELSA' to 'RRI'. *Life Sciences Society and Policy* 10(1): 11.

Chapter 2
Promises, Expectations and Visions: On Appraising the Plausibility of Socio-Technical Futures

> *We are like sailors who have to rebuild their ship on the open sea, without ever being able to dismantle it in dry dock and reconstruct it from the best components (Neurath O, 1932 [1983] p 92)*

Abstract When a scientific and technological field is still emerging, promises of its social desirability and warnings about its potential negative effects are wide spread. The dawn of the Human Genome Project (HGP) is an exemplar in this respect. The high expectations that emerged from the early stage of genomics research have been drastically deflated while the field has continued to develop. Difficulties, uncertainty and unanticipated constraints arose at later stages of research to challenge the initial expectations of scientists, investors, policy makers, clinicians, patients and other social groups. These projections into the future described an individual or collective belief in the possibility that a certain state of affairs would come into being. Drawing on a diverse set of literature, this chapter discusses the strategic, performative and normative character of visions of technological futures. It argues that if, on the one hand, visions are morally characterized as implicitly normative, while, on the other, technologies and our morality mutual shape each other, the analysis of their "plausibility", rather than their "desirability", becomes crucial. The chapter concludes by outlining an approach to the reconstruction of plausible expectations around emerging technologies consisting of *thickening*, *zooming in/out* and *situating* visions of emerging technologies.

Keywords Expectations • Visions • Plausibility • Vision assessment • Philosophy of technology • STS

2.1 Expecting Future Science and Technologies

When a scientific and technological field is still emerging, promises of its social desirability and warnings about its potential negative effects are wide spread. The dawn of the Human Genome Project (HGP) is an exemplar in this respect. In the early 1990s, the hope of mapping the entire human genome by 2005 sustained the

American Congress' decision of allocating $3 billion to this goal. In June 1998, when the working draft of the human genome sequence was announced by Francis Collins, the director of the HGP, the enthusiasm for mapping, sequencing and determining the function of human DNA was so high that the project was promised to end 2 years before the expected date (Collins et al. 1998). In the often quoted Shattuck lecture on the "Medical and societal consequences of the Human Genome Project" the HGP director describes in a prophetic tone how "the history of biology was forever altered" by a "bold" decision to initiate the enterprise that had the goal of characterizing "in ultimate detail the complete set of genetic instructions of the human being" and concluding that "the genetic revolution in medicine is under way"

> Once the contributing genes and their disease predisposing variants have been identified, diagnostic tests can be developed to predict future risk — but these tests are most effective when a preventive strategy is available to reduce the risk in persons found to be predisposed to a particular disease. Another rapidly developing application of diagnostics is pharmacogenomics, the prediction of responsiveness to drugs. Ultimately, the real payoff of genetic research will be the development of new gene therapies and drug therapies, but they will generally require many more years of intensive research. (Collins 1999)

At the end of his lecture, after explaining the repercussions of the HGP in medicine and the ethical, legal and social issues that could arise from them, Collins provides a short narration, a story or, as he calls it, a "scenario", describing medical practice in 2010. This scenario depicts a fictional character, Mr. John, testing his DNA and receiving detailed information about risk association and disease predisposition and explains how Mr. John decides to act upon it (Collins even provides a table showing what this information would look like).[1]

Collins' vision was not an isolated case at the beginning of 2000. Many voices rose from the fields of genetics supporting hopes of the medical opportunities associated with DNA research. Not only could genetic research be applied to pre-natal and postnatal screening to identify babies at risk of developing certain diseases, but it could also lead to personalized medicine and better-tailored preventive medicine. Chronic diseases could be managed, increased risk identified and preventive measures taken. Furthermore, the individual susceptibility to common disorders could be determined. Some also pointed out the enormous potential for a new field of pharmacogenomics, in which not only diagnoses, but also treatments could be personalized by determining individual susceptibility to drugs (Epstein 2004; Richards 2001). These hopes were accompanied by worries about potential ethical and social implications. Such fears were partially addressed by allocating 5 % of the annual budget of the HGP to

---

[1] In Collins' words: "As genome technology moves from the laboratory to the health care setting, new methods will make it possible to read the instructions contained in an individual person's DNA. Such knowledge may foretell future disease and alert patients and their health care providers to undertake better preventive strategies. In the wrong hands, however, that same information could be used to discriminate against or stigmatize a person. In response to this concern, the Human Genome Project has catalyzed the development of policy options for lawmakers to consider in their efforts to prohibit genetic discrimination and to protect the privacy of genetic information. The stage is set to solve these vexing problems with effective federal legislation, but this window of opportunity will not stay open indefinitely" (Collins 1999).

exploring the Ethical, Legal and Social Issues (ELSI) related to this scientific enterprise.[2] ELSI research focused mainly on questions of data privacy, informed consent, inequality and stigmatization induced by the association of genes with behavioral traits and social or racial differences.[3] These studies addressed these ethical and social implications of wide spread use of genetics in healthcare. Independent from the question of whether these implications were appraised as positive or negative, these studies took for granted that the promises on genomics would be fulfilled.

The enthusiasm of the promoters of the project was balanced by many "skeptics" who doubted whether genetics would actually revolutionize medicine (Richards 2001). According to these skeptics, mapping and sequencing the human genome could only lead to the identification of genes causing monogenetic diseases, but this would benefit only a small percentage of the population (Holtzman and Marteau 2000: 141). Furthermore, effective treatment would lag behind. In the following years, several studies confirmed the skeptics' positions. For example, the hope that the association between a gene variant and a certain disorder could lead to a diagnostic test to identify people who carry the variant gene was soon disappointed by scientific evidence. In fact, common multifactorial disorders are determined by "complex gene-gene and gene-environment interactions" (Janssens and Khoury 2006) that are not captured by generic risk profiles and require increasing reference to susceptibility and risks estimates.[4] Also in the case of monogenetic disorders, like Huntington's disease and breast cancer, evidence has shown that (1) there might be not only one, but more genes associated with the disease (for example BRCA 1 and 2 in the case of breast cancer) and (2) when the associated risk is estimated on a larger population scale (and not only in families with a high incidence of the disease), the associated risk estimate is much lower and therefore less informative (Burke et al. 2002). Finally, for pre-natal tests, where the predictability of predisposition seemed quite reliable, studies have shown that prospective patients and clinicians often had to deal with uncertain "grey results" that complicated rather than facilitated clinical decisions (van Zwieten 2008). The high expectations from the early stage of genomics research have been drastically deflated while the field further developed. Difficulties, uncertainty and unanticipated constraints arose at later stages of research to challenge the initial expectations of scientists, investors, policy makers, clinicians, patients and other social groups. The initial "genohype" (Holtzman 1999) has

---

[2] ELSI funding programs were presented in Chap. 1 as an institutional trend towards the "outsourcing" of ethical reflection to universities and research institutes (Sect. 1.3.3).

[3] For more information on the type of ELSI projects conducted within the HGP see http://www.ornl.gov/sci/techresources/Human_Genome/elsi/elsi.shtml

[4] On the basis of a simulation study–conducted to evaluate the predictive value and inheritance patterns of generic profiles and comparing the results with genetic tests for Huntington's disease and hereditary cancers – (Janssens and Khoury 2006) shows how actual data on risk estimate in genetic variation carriers are only slightly higher than that in non-carriers. Even if research has been conducted in improving multiple genetic testing (genomic profiling), neither the predictions provided by these tests are informative, since "most of individuals have disease risks that are only slightly higher or lower than the average disease risk in the population".

required re-consideration in light of the little progress that has been achieved in translating scientific advances to the clinic, let alone into health gains.

ELSI research on genomics was also assessed as hyped. Holtzman introduces the term 'ethereal debates' to point out that the ethical, legal and social issues about behavioral genetics and the possibility of genetic enhancement "are built on a house of cards".[5] ELSI research on genomics had (and has) the mandate of exploring the moral desirability of the futures proposed by scientists and proponents of the genomic revolution. However, in doing so, this research is bound by scientists' hopes and expectations and might even contribute to the genohype (see also Hedgecoe 2010 and discussion in Sect. 1.4.2). As the case of the Human Genome Project suggests, ethical reflection on emerging technologies has to deal with uncertain objects: expectations concerning new science and technology; promises concerning how emerging science will create better futures; hopes of innovative solutions to current problems. Expectations of genomics are diverse and have spread to different social actors, including not only molecular biologists, chemists, medical geneticists, but also policy makers, pharmaceutical companies, mass media, insurance companies, diagnostic laboratories, and prospective parents. Each social actor holds different expectations, hopes, visions and projections about the future of genetics in the medical practice. These projections into the future describe an individual or collective belief in the possibility that a certain state of affairs will come into being.

An upstream ethical debate on emerging technologies based on hype may indeed result in "ethereal debates". By inquiring and reflecting on issues such as privacy or discrimination in the field of genomics, bioethicists intentionally or unintentionally presuppose that a certain technological development will come into being. Chapter 1 has already introduced criticism against speculation in ethics and the risk for ELSI studies of addressing questions that go beyond the limits of technical possibilities. According to this criticism, ethical debate should start by assessing the quality of expectations on emerging science and technologies, a "reality check" (Nordmann and Rip 2009, see Sect. 1.5). The rest of the chapter will discuss insights from the scientific and philosophical literature that may help to develop a methodology for such an assessment. In order to do so it is crucial to understand what kind of objects are expectations concerning emerging scientific and technological enterprises. Section 2.2 discusses recent studies in the sociology of expectations that explain why expectations surrounding emerging technologies are more than just beliefs or representations of future technological worlds. This literature points out the socially constructed nature of expectations and their strategic and performative role. Section 2.3 introduces the literature on Leitbilder and vision assessment. It emphasizes that visions of future technologies hold a normative content that is not always explicit and therefore should be disclosed. In Sect. 2.4, insights from empirical philosophy of technology illustrate how technologies often do much more, and very different things, than what they were supposed to achieve. Technologies engage human

---

[5] This quote is taken from an interview with Neal Holtzmann, director of genetics and public policy at Johns Hopkins University, a skeptical voice during the debate on the revolution of genomics in medicine (in Richards 2001).

beings in a rich array of relationships. The final section of the chapter (Sect. 2.5) recollects what can be learned from this literature review. I argue that it is indeed important to assess the desirability of emerging technologies, but that we should not take for granted claims that a technology will realize certain values. We need to analyze the *plausibility* of such claims. I propose that such an analysis could proceed by reconstructing and assessing three dimensions of specific expectations: the expected artifact, its use and its value.

2.2 The Social Construction of the Future

In the last 20 years, a number of studies have investigated how expectations play a strategic role in the dynamics of innovation (van Lente 1993; Brown et al. 2000; Hedgecoe and Martin 2003; Michael and Brown 2003; Merkerk and Robinson 2006; Selin 2007). In these studies, at the interface between the tradition of social-constructivism (Bijker et al. 1987) and innovation studies, it emerges that expectations guide actors' decisions, legitimize their actions, attract interest or raise worries (Borup et al. 2006).

Van Lente (1993), in a pioneering study, shows that expectation statements circulate at different levels and stages of the technological innovation process with a number of functions. Expectations at the "macro-level" are general *promises* concerning the technology, "broad, diffuse, general expectations, which have the form of scenarios" (*ibid.*: 184). An example of this type of promise that van Lente provides is: 'Memory chips will be of growing importance in all kinds of industry'. These macro-level expectations are the ones most visible for lay people, and probably also for ethicists. They are used by spokespersons to legitimize certain techno-scientific domains when the social importance of a technology is not yet taken for granted and needs to be justified. Such expectations outline scenarios of technological and societal trends and are usually considered by social actors as "taken for granted background" (*ibid.*: 184) rather than as issues for discussion.

At a "micro-level", expectations circulate among experts in a laboratory, in which they are expressed as quantitative *specifications* about future artifacts. These "search-expectations" "contain specifications for the artifacts, systems, or processes to be developed." One example of micro-level expectation is: 'The breakthrough value of Tenax paper will be 50 % better'" (*ibid.*: 181). These expectations are used as "heuristics" or "guiding search processes" and set requirements for the local agenda, for example, within a research group.

At an intermediate (meso-)level, expectations are qualitative statements about the *functions* that the technology will presumably fulfill (*ibid.*: 182). They are general statements used by actors outside their own research context to position themselves within a certain scientific or technological field, but also to show the opportunities of a domain (e.g. "cleaning waste water has great potential") and guide decisions for funding allocation. The boundaries between the macro, meso and micro level are fuzzy. Van Lente stresses that statements, agendas and actions among different levels are "nested": the acceptance of a broader promise requires the acceptance of other expectations associated with it.

Van Lente's multi-level analysis points out that expectations can have different forms and functions depending on the actors, contexts, and time. Other authors emphasize that they can be "inscribed" in actions, material, and machines (Borup et al. 2006; Brown 2000). Material artifacts, practical decisions, and courses of action embody expectations. Therefore, beyond the explicit discourses, expectations manifest themselves also in these material practices. To use the example of the Human Genome Project, expectations that genetic science has a special role in healthcare materialized in research proposals, conferences, funding programs, policy decisions and ethical reports. These statements about technological futures are not (only) epistemic constructs on future state of affairs, representing a belief in a state of affairs. Expectations "do something" else, they advise, show direction and create obligations (van Lente 1993: 190). Actors in technological development use expectations as a "resource" when they need to "legitimize" their arguments and actions. They can also be used as a resource to convince other actors to act, for example to attract the interest of investors, to mobilize funds or to achieve the consensus of public opinion. By mobilizing support and ensuring funding, expectations play an important role in the innovation process, because they allow the development of protected spaces for an emerging techno-scientific field to grow.

Therefore, expectations are *performative*: they are not merely *representations* of the future but they *do something* (van Lente 1993; Brown 2000). They allocate roles to social actors: for example, Collins' expectations that the genetic revolution would happen attributed a role to pharmaceutical companies in order to encourage them to invest into pharmacogenomics or to the American Congress to allocate money for research in the Human Genome Project. In Collins' expectations, different social actors (biologists, politicians, pharmaceutical companies, etc.) performed a role in view of the success of the promise of the "genetic revolution". Expectations create a "promise-requirement cycle" (Brown 2000; Geels and Smits 2000), in which protected spaces (or niches) for innovative technology are created (Konrad 2006), stakeholders position themselves and others in a certain techno-social system, and funds are allocated. Once these expectations are shared they acquire autonomous force and set demands on actors.[6] These expectations are not produced by an intentionally well-designed overarching plan or by a determined chain of necessary steps; instead they take shape through a contingent, bottom-up, decentralized process. The competition among heterogeneous visions also allows space to discuss, negotiate and construct new configurations (Arnaldi 2010).

Expectations not only vary "vertically" with respect to the level of the innovation process, they also vary "horizontally" among different actors. In the case of early expectations, the degree of involvement of actors in technological development or

---

[6] They have an active role in creating agendas and in "interlocking" activities. Promises are, in fact, taken up in the agenda and allow the creation of new interactions and new roles for stakeholders, who gather around the promise. The interlocking of these positions creates a requirement that demands some action. In this way, expectations create obligations among the stakeholders and establish objectives that become constraining: they create *emerging irreversibilities* (Merkerk and Robinson 2006).

scientific practice influences their assessment of the feasibility of such scenarios. Expectations appear more authoritative to those who exert little influence over the outcome of a promise (a broader public, for example). Researchers conducting research, in contrast, are more aware of the uncertainties and challenges involved (Michael and Brown 2003).[7]

The strategic and performative role of expectations in innovation dynamics suggests that they are more than just experts' statements on the future: they are rhetorical constructions (van Lente 1993). Scenarios and promises have a legitimizing role for new and not-yet established fields, as they may be used to convince an audience to do something: for example, funding a research proposal, lobbying for a research field in a policy making environment, welcoming a future technology in a clinical practice, or using such technology. The use of rhetorical strategies is crucial especially for researchers who deal with issues that are unfamiliar to the audience: rhetorical tools are then mobilized to convince the audience that they are dealing with something real. Promotional rhetoric (Guice 1999), "hope" discourse (Brown 2006: Moreira and Palladino 2005), and hype (Michael and Brown 2003) have often been highlighted as a means to persuade the audience of scientific peers, funders and the larger public by appealing to their emotions.

To sum up, the literature from the Science Technology and Innovation studies shows, first, that there are different types of projections of images of the future, with a different role and function in the process of technological development. It further points out their *vertical* variability across the levels and stages of technology development, their *horizontal* variability across actors and contexts, their *performative* character in directing actions and creating requirements, and their *strategic* and *rhetorical* component. These aspects all have to be taken into account when reflecting on the ethical desirability of emerging technologies. As highlighted in Chap. 1, some of these studies around expectations have been developed within some second generation approach to Technology Assessment, especially Constructive Technology Assessment (see Sect. 1.2, Merkerk and Robinson 2006; Merkerk 2007). These approaches develop further understandings of how expectations play a role in stakeholders' interactions and feed this learning back into stakeholders' workshops.

2.3 The Guiding Normativity in Technological Visions

Promises of technological futures have also been studied, with respect to their normative content, as *visions* that carry and promote an ideal of a *desirable* world. In particular, macro-level expectations or promises have been addressed by German authors as *Leitbilder* or guiding images (Dierkes et al. 1996; Grin and Grunwald 2000; Quist 2007). These "images" of attainable futures are shared by different

---

[7] The analysis of these dynamics also illustrates a temporal dimension of techno-social expectations (Borup et al. 2006). They change over time: while early on in the innovation journey promises are essential to attract attention and create niches for innovation to develop, later on these promises will likely be disappointed, giving way to disillusionment (Michael and Brown 2003).

actors and "guide" their action and interactions (Grin 2000). Rather than focusing on the dynamics of interactions in the innovation process, studies on *Leitbilder* emphasize and analyze their normative content. *Futuribles* (Grin 2000) or visions of desirable futures carry the values, worldviews, and deep preferences of the holders. A vision is "the world described by someone who is asked why particular technologies are desirable" (Grin and Grunwald 2000: 53). Take as an example Collins' promising scenarios on how genetics will revolutionize medicine. Why is genetics desirable? Because it will provide diagnostic tests and these tests will offer knowledge that "alerts patients" and offer care providers "better preventive strategies". These consequences are presented as good and valuable for society. The literature on "visions" suggests that expectations are not only projections of how a technology or a scientific discovery will play a role in a *social* context; they also imply a moral judgement on that future world.

The moral character of these scenarios can be understood when we think of the literary tradition of utopias and dystopias. This tradition has played an important role not only in modulating public debates, but also in framing scientific research within a particular universe of meaning. For example, Aldous Huxley's novel *Brave New World* (1969 [1932]) provides a tale of how the utopia of a healthier society implied in technologies for pre-natal genetic screening might turn into a dystopia of social re-organization. This dystopian turn questions the desirability of the quality of life, human relationships and self-perceptions resulting from these technologies, as well as the quality of democratic processes. The novel describes the values implicit in utopian visions on pre-natal genetic screening and explores the (un) desirability of its worldview. In doing so, it offers a scenario that has a different moral connotation from the one offered by Collins, about the desirable future of medicine in the year 2010. Different though they are, the comparison shows that both visions are value laden. Collins' vision of how the patient in 2010 manages his health through genetic screening and adopts preventive measures to balance his risks is not a neutral description of the future. Instead, it is a wishful image of a desirable state of affairs, in which every individual is in control of her health condition.

Visions are rooted in culture, its morality and traditions. In this sense, they are stable or "culturally coherent" (Grin and Grunwald 2000). Visions relate the past to the future, acting as carriers of cultural repertoires and shared meanings. In her analysis of 35 conversations with nano-research scientists and engineers, Berne (2006) explains that some "old myths", like the idea of mastering the physical universe or the perpetual self-improvement, dominate the discourse of nanotechnologists. These old *archetypes* are drawn upon because they provide descriptions of the place of human beings in the world that still seems viable in our technological culture. In this way they support the technological innovation. These visions are vehicles for interaction and communication among actors since they carry some general meanings allegedly shared by different social actors (Dierkes et al. 1996). However, because of their general and suggestive reference to a shared cultural repertoire, these visions can be carriers of different meanings and interpretations among different stakeholders.

2.3 The Guiding Normativity in Technological Visions

At the same time, visions are transformative and challenging because they propose alternative states of affairs from the current ones and guide technology developers, policy makers, industrialists and the larger public towards such alternative scenarios (Grin and Grunwald 2000: 176). A scenario of an alternative future motivates actors to act in order to realize it, or to raise opposition in order to prevent it from happening. In fact, future scenarios described by scientists, media or science fiction writers inspired by new technologies play a role in social actors' decisions at different levels. For example, the high impact images of Crichton's novel *Prey* (2002), depicting predator nanorobots escaped from human control, are mentioned in policy documents about nanotechnology (Selin 2007: 205). The tension between "old" and "new" in the dialogue between past, present and future emerges also in the ambivalent discourses of actors who selectively emphasize the continuity of their innovation with the past or its novel purport (Swierstra and Rip 2007; Shelley-Egan 2011).[8]

The concept of "leitbilder" emphasizes that visions of the future are intrinsically moral; they describe a desired world and a future state of affairs within a specific value framework. They are also *normative* in the sense that they guide towards these desirable worlds. Although this normative content of visions clearly appears in the literary genre of Science Fiction as utopia or dystopia, it remains often implicit in technology developers' guiding visions. Even if preferences, worldviews and value systems are often unspoken in technological visions, it does not mean that they do not play a guiding role in decision-making processes. It is legitimate to question, then, what is the normative content of these visions. The emphasis on the implicit normative content of these guiding images adds another aspect to the discussion on the desirability of emerging technologies addressed in Chap. 1. When engaging in an assessment of emerging technologies that aims at an exploration of moral issues and a normative discussion, it seems that not only the strategic and contextual role of expectations should be taken into account, analyzed and fed back to stakeholders (as CTA already does), but also their normative content equally should be considered. This is the position held by the authors emphasizing that one of the missions of Technology Assessment should be to facilitate a democratic discussion around the normative content of expectations about emerging technologies. As suggested by Grin and Grunwald (2000), a *Vision Assessment* approach should explore and articulate the normative meaning of visions that is often left unspoken.

Grin and Grunwald (2000) make a case for vision assessment. The authors point out that the Dutch Investigative Medicine Program assessed cochlear implants for deaf people from a (dominant) clinical perspective according to which deafness is a handicap. Deaf Organizations critically responded to such an assessment, pointing out that "the perspective on deafness as a handicap is not shared by all" (*ibid.*: 58). Therefore, the authors argue that an assessment of new and emerging technologies should uncover assumptions and perspectives in technological development that are presented as universal, but are instead shared only by some stakeholders. Instead of

---

[8] Selin (2007) shows how nanoscientists know both how and when to mobilize Drexler's visions on nanotechnology, and how and when to dismiss them as science fiction.

taking the dominant perspective, technology assessors have to include the perspectives and visions of all the relevant stakeholders in the deliberation process, bringing their voices in, creating spaces for confronting, communicating, arguing with other stakeholders on questions of what is important. The authors therefore address the question of how technology assessment can help to "uncover and critically examine visions, expectations and perspectives underlying the development and use of the technology" (p. 62). With "vision assessment", they propose an interactive approach that consists of broad consultations with the expected users of the technology and compare them with the visions of the technology developers'. In these exercises, stakeholders are asked to debate standards and criteria of merit of the technology, to discuss problem definitions and possible solutions, to envision the contexts of use of the technology, and describe the "desirable final state". This discussion should aim to make explicit the different needs of various stakeholders, considering them to be legitimate and scrutinizing them towards reaching some form of agreement.

The Vision Assessment approach addresses the "normative deficit" diagnosed in TA (see Chap. 1) by proposing to disclose and discuss the normative content of visions on emerging technologies. According to this approach, "expectations for the future reflect the values, worldviews and deep preferences of those who hold them" (Grin and Grunwald 2000: 11) and they contain assumptions about what is a "desirable" world. In the preface, I referred to the etymology of the word "assess" and explained how it is related to the idea of "evaluating" something or "comparing it with some values". The vision assessment approach points out the need for making the values that guide the assessment of emerging technologies explicit and for discussing them among stakeholders, rather than taking them for granted. This is a crucial point in that it broadens the discussion of expectations. Not only does it elucidate the rhetorical and strategic role of expectations in guiding them towards *normatively* connoted goals, but it also offers some analytic tools to disclose the implicit normative assumptions intrinsic to actors' expectations (see Chap. 1). Accordingly, normative assumptions hidden behind metaphors, vague concepts and criteria of desirability are opened up and critically addressed. This approach has the merit of claiming a space for a discussion about the value assumptions and worldviews of various stakeholders' expectations concerning the desirability of certain technologies. Moral values, however, can be explored beyond deliberative exercises that articulate stakeholders' visions, beliefs and framing of what is a desirable world. As studies in the fields of STS and Philosophy of Technology have frequently shown, technologies have moral and political connotations (or even agency), and values are "inscribed" in artifacts. It is not only a matter of how moral, cultural or political framings shape and direct visions of technologies, but also how they shape the very materiality of the artifact. The next section introduces some philosophical considerations concerning the interaction between technology and morality. This reflection carried out in the broad disciplinary field of philosophy of technology broadens and enriches our expectations concerning the "good consequences of technologies": technologies are not simply tools or a means to an end they also embody some values, direct human actions, and alter human practices and values.

2.4 Beyond an Instrumentalist View: Technology and Morality

During the first half of the twentieth century several philosophers reflected on the role and essence of technology. Some of these philosophers like Jacques Ellul, Karl Jaspers, Martin Heidegger, and Gunter Anders engaged in transcendental reflections on the conditions of the possibility of technology. "Technology" with a capital T was often considered either as a neutral means to an end – what is often referred to as "instrumental" view – or as an autonomous entity in stark contrast with the human sphere of meaning and culture – also called a "deterministic" view. Towards the end of the 1990s, European and American philosophers of technology moved away from this attitude (Kroes and Meijers 2001: Achterhuis 2001). Several contemporary philosophers of technology – such as Andrew Feenberg, Alfred Borgmann, Don Ihde, Carl Mitcham, and Joseph Pitt – have reflected on the concrete empirical manifestations of different technologies, moving the focus from the meaning and consequences of Technology, as an abstract entity, to actual technologies: diagnostic devices, communication technologies, scientific or medical instruments, or virtual realities. These become the new focus of philosophical inquiry, implying a different conceptualization of the instrumental and deterministic views. Technology becomes a human-controlled tool to reach a pre-determined end.[9]

Many authors in contemporary philosophy of technology, drawing on Science and Technology Studies, defend an anti-instrumentalist conception of technology. They argue that artifacts are not simple tools or extensions of human organs. On the contrary, "there is a flux of possibilities that you suddenly envisage when you handle a hammer" (Latour and Venn 2002: 250). This is exemplified in Stanley Kubrick's *2001*, when the monkey comes to realize that a dinosaur jawbone, initially used as a hammer, can also be used as a club to kill. In this sense, technologies "are not innocent". Their use *translates* or *displaces* our initial goal because "*we have changed the end in changing the means*" (*ibidem*: 253). In *mediating* our actions, artifacts change our goals, in the same way the installation of speed bumps in the street does not simply fulfill the function of calming the traffic. Instead, their introduction in towns and villages initiates a complex network of social interactions in which drivers, police agents and city halls are

---

[9] Philosophy of technology is a diverse and broad scholarly field that is hard to define in a systematic way. In a way, even Plato's myth at the opening of Chap. 1 can be considered as a philosophical reflection on the role of artifacts and technologies. In the last 20 years some self-proclaimed philosophers of technology have drawn largely on sociological studies of science and technology (or Science and Technology Studies, STS) making the attempt to define the disciplinary field difficult, if not pointless. Furthermore, it is hard to draw a clear-cut distinction between philosophical reflection on scientific method, activity and role from philosophical reflection on technology. It goes beyond the scope of this book to provide a fair description of the field of philosophy of technology that would do justice to the wide variety of authors who have contributed to this field. In particular, Mitcham (1994) distinguishes a "humanities philosophy of technology", which is continuous with social sciences and humanities, as opposed to an "analytic philosophy of technology" that seeks continuity with philosophy of science. (http://plato.stanford.edu/entries/technology/)

involved. Technological systems are "opaque black boxes" in the way that goals, intentions and outcomes are connected.

The post-phenomenological approach to technology (Ihde 1991) has emphasized that technologies are part of the human condition. They are embedded in the way we perceive the world, similar to how a pair of glasses or a thermometer help us in relating to the external reality and constituting our own. Technologies mediate both our perceptions and actions (Verbeek 2005). This is clear, for example, in the case of the monitoring device for blood sugar measurement. When this device was introduced in the routine of diabetic patients, it didn't simply fulfill a function of measuring the level of sugar in the blood, but also it created the practice of self-monitoring. This new practice implied not only different actions, behaviors and routines, but also new standards of normality and different relations between patients and their own body (Mol 2000).

Technologies have a political dimension because they shape and determine power-relations. Langdon Winner asks, for example, if artifacts *have politics*. His answer is that they do. Robert Moses' design of bridges over the Long Island Parkway, for example, was intentionally low to prevent buses and therefore the poor black minority to pass under it. Similarly, nuclear power plants presuppose a hierarchical political structure for their security management. Andrew Feenberg gives another example. He explains how the controlling ideology embedded in the Minitel technology in France (a precursor Internet) was reshaped by the users who – opposing the intent of the designers and policy makers – adopted it as an instant messaging device (1995). Feenberg welcomes user creativity as a democratic force, as opposed to an authoritarian technocratic design. This example shows that it is possible to democratize technologies, a task Feenberg prescribes to philosophers of technology.

The examples of the speed bumps, the blood sugar meter and the Long Island Parkway's bridges show that artifacts are designed to make people behave in some particular way. The speed bumps *prescribe* drivers to slow down; the blood sugar meter *requires* diabetic patients to monitor the level of glucose several times a day; the Long Island Parkway's bridges *forbid* non-car owners to reach Long Island. In this sense, technologies are moral agents because they discipline people's behavior and distribute roles among actors and stakeholders (Akrich 1992; Latour 1992).

The notion of "script", articulated in the context of the Actor-Network Theory by Bruno Latour and Madeleine Akrich, captures the idea that "like a film script, technical objects define a framework of action together with the actors and space in which they are supposed to act". For example, the cumbersome shape and weight of some hotel key-chains encourage the users to return the room key to reception before leaving the hotel (Akrich and Latour 1992). The key-chain contains a "program of action" inscribed in it (Latour 1992): a non-written instruction of how it should be used and by whom. While defining their technical objects, designers make assumptions about the world in which these will be inserted. Technologists' visions of the world are "inscribed" in the new object.

2.4 Beyond an Instrumentalist View: Technology and Morality

Designers thus define actors with specific tastes, competences, motives, aspirations, political prejudices, and the rest, and they assume that morality, technology, science, and economy will evolve in particular ways (*ibidem*, 208)

Thus, on the one hand, the "composition" of a technical object constrains the social actors[10] and their mutual relationships and, on the other hand, these actors reshape the object and its use. The design process, by defining the components of the artifact, also defines the relationships among actors. Indeed artifacts *pre-scribe*, oblige, allow or forbid specific behaviors and distribute roles and responsibilities among users.

The concept of "script" emphasizes how designers unintentionally inscribe a moral agency in the artifacts. The way the designers imagine and determine the user in the technical object, her relations, and the context of use, may be quite different from the actual reality. Akrich emphasizes that "the world inscribed in the object" might be different from "the world described by its displacement". The script analysis suggested by Akrich and Latour consists in providing "thick descriptions" of the artifact design and the artifact in a social practice. The comparison between these two descriptions shows the mismatches and sketches the *geography of responsibilities* around technical artifacts.

Along the same lines, philosophers in computer ethics have argued that values are "embodied" or "built" into artifacts. The "embedded values" approach (Nissenbaum 1998) "holds that computer systems and software are not morally neutral and that it is possible to identify tendencies in them to promote or demote particular moral values and norms" (Brey 2009: 42). Computer systems present built-in moral consequences that are inherent in their design: for example, users' autonomy is undermined by software agents that hide relevant information from users (Friedman and Nissenbaum 1997). The use of technological artifacts can also promote (or demote) the realization of moral values, such as privacy. This means that artifacts are value-laden and design can be "value sensitive" (Friedman et al. 2003): designers can intentionally inscribe values in artifacts, shaping them, in order to support or undermine enduring human values.

Technologies do not only impose a conduct on their users. Technologies interact with our morality at a broader level. Swierstra (2010) remarks scholars' inclination for a one-way analysis emphasizing either the adaptation of morality to technology or the moral and normative construction of technologies. He argues that the interplay of technology and morality should be understood in a symmetric way. This non-linearity of techno-moral dynamics is emphasized in the metaphor of morality as a "force field" in which different norms and values stand in relation to each other. In the force field of morality, different vaguely defined values are in tension with one another. These tensions are usually solved through compromise, by creating a

---

[10] Latour 1987 introduces the concept of "actant" borrowing it from the field of semiotics. It refers to human and non-human entities whose agency is mediated by a spokesperson. Since for the purpose of this study I will not engage in a thorough semiotic analysis, I will keep the "actor" vocabulary to avoid unnecessary complex jargon. For a definition of the term "actant" see also Akrich and Latour 1992.

hierarchy among values, or by distributing them in different spheres of application. The force of these values can only be assessed relative to other values, i.e. when an individual is forced to choose between two values. In these cases, "one norm pulls harder than the other".

> Morality can best be understood as a force field wherein conflicting norms and values compete for hegemony. In this competition, technology can sometimes tip the scales. The [birth control] pill created new conditions that enabled more people than previously to take the dominant norms and values less seriously. At the same time, previously marginalized norms and values now come to the fore and gain societal acceptability. (*ibidem*, my translation)

To clarify this last point we can use the example of the telephone, which was first sold as a business device. Even when they reached private households, they were only for functional use, for example, so that (house)wives could order groceries. Only later did the telephone become a device for social interaction (despite telephone companies at the time trying to hinder this "improper" use) (Fischer 1992, 2011). Practices and routines have been profoundly changed by the telephone, with cell phones adding further changes by reconnecting people to their smaller world everywhere and at any time (Humphreys 2005; Goggin 2006). Within these societal and cultural changes, Pamela Pauls explains in the *New York Times* how people have started a new habit of not picking up the phone so that they remain undisturbed by a phone call (Pauls 2011). Calling somebody without emailing first is felt progressively to be "rude" or "awkward" while telephone companies register a decrease in the voice spending, while text and surf spending increases (Ofcom 2012). When making a phone call people were perceived to be intruding into their friends', families' and (potential) customers' lives. This threat to the value of "privacy" is exacerbated with the use of mobile phones that allow another to invade a person's private life anywhere and at any time. In the vocabulary of techno-moral dynamics, the value of "sociability" associated with the telephone seems to be in conflict with the values of "privacy" and respect of other people's private space. This change in habits and etiquette goes alongside a different priority relation between the value of sociability and privacy in the context of calling: the privacy value "pulls harder" than the sociability value. This exemplifies a position in which technologies are not only shaped by societal values and ideas of "good" but they also mediate our actions and redistribute responsibility, affect our routines, perceptions, standards, concepts, and influence which values we prioritize in the "force field" of morality and how we interpret them. Technology and morality mutually shape one another (Swierstra 2010, 2013).

The literature from empirical philosophy of technology, then, shows the inadequacy of the view that technology is simply an instrument to achieve a fixed goal. Because of its material hermeneutics and agency, an emerging device can be expected to engage in a complex array of relationships with its users. Within these relationships some values or power-relations are promoted more than others and roles and responsibilities are distributed in a specific way. Values and assumptions about the user and the context of use are inscribed into the design of artifacts. Furthermore, technologies modulate what we value and how much we value it.

When an emerging technology is described as a means to achieve specific values, as most expectations do, these aspects are neglected. Furthermore, this intrinsic normativity in the material shaping of technological artifacts is often neglected in accounts that focus on stakeholders' value propositions and visions. The interaction between technology and morality contradicts the instrumentalist expectations that a certain technology is just a neutral means to achieve something that is (pre)determined by its user. Before questioning whether the values associated with an emerging technology are desirable, then, we need to ask how *plausible* it is that an emerging technology will instrumentally and linearly produce those (un)valuable consequences.

2.5 Analyzing Expectations' *Plausibility*: A Proposal

Now we can go back to the question outlined at the end of Chap. 1: how to integrate ethical inquiry into Technology Assessment (TA) in a way that we can find a balance between, on the one hand, the need to explore the moral dimension of technological innovation and engage in normative discussions about acceptability of these technologies and, on the other hand, the danger of engaging in highly speculative and "ethereal" debates on the moral significance of new technologies and their consequences on our societal or moral apparatus (values, concepts, etc.)? We have seen in this chapter how the sociology of expectations points out that prospective statements about the functioning of emerging technologies in societal contexts are not objective representations of scientific facts, but projections of broad techno-social configurations. They have a strategic role of protecting innovations by securing funding and support. The context of utterance, the actors involved, their positioning and interest should be taken into account when addressing expectations on emerging technologies. Their rhetorical and strategic character, however, does not mean that expectations should be discarded all together as objects of ethical discussion: the performative character of expectations means that they "do" things independently from their accuracy or reliability. As discussed, promises concerning emerging technologies are visions of desirable futures. They have a normative content that guides the actors' decisions in the innovation process, justifies their choices, and mediates their communications. Actors' projections of the future are rooted in a past of established traditions, norms and "myths" which legitimize and give meanings to them. As claimed by some authors, these visions are a rich resource that should be assessed: their implicit values, preferences and worldviews should be made explicit and their desirability openly discussed in democratic and deliberative settings, wherein what is "good", according to some stakeholders, may appear less so to others. How to foster an explorative and yet grounded discussion on emerging technologies when the object of analysis are expectations that are, as we have seen in this chapter, strategic, rhetorically constructed, and normative loaded? How to do so, while acknowledging that the technology and morality mutually shape one another?

2.5.1 Desirability Versus Plausibility

Although the Vision Assessment approach addresses the crucial issue of exploring normative stances in stakeholders' visions, there are two issues that need attention. First, values and worldviews are not only held by stakeholders *about* technologies, but also inscribed *in* technological artifacts. As studies in philosophy of technology and STS have shown, technologies materially mediate human agency and identity and also shape – while being shaped by – moral values. This morality in the technology calls for a close look beyond stakeholders' beliefs and values, into material practices of technology development. Interestingly, expectations that a technology will have some valuable social outcome do not always take into account the changing aspect of the normative framework against which such desirability is assessed. Second, "assessing visions" requires not only making their normative content explicit and fostering democratic discussions among stakeholders about the *normative* acceptability of such visions, but it also entails making explicit and discussing the *epistemic* justification of such normative content. If the vision is "the world described by someone who is asked why particular technologies are desirable" (Grin and Grunwald 2000: 53), then, at being posed the question "why is genetics desirable?" Francis Collins will likely answer that it is so because it will provide diagnostic tests that will enable preventive care of pre-symptomatic individuals. These, according to him, are good and valuable outcomes. The vision assessment approach aims at disclosing and assessing the desirability (D) of the vision that "genomic-based healthcare is good". The *desirability*-question asks why these outcomes are good and valuable, according to which conceptions of good and based on which values. Such questions probe stakeholders' answers and enable them to compare their normative stances. However, the visions of genomics (more or less) implicitly suggest that not only a genomic-based healthcare is good, but also that there are good reasons to believe that it will be realized in practice. However, if technological artifacts embody some values, promote or forbid actions, distribute responsibilities and are in a mutually shaping relationship with our value system, it is unlikely that a certain technology will "simply" do something good. Given the current technical, economic, social, cultural AND moral constraints, can genetics research indeed be expected to produce these good consequences? This is what I would call a *plausibility* (P) question. Assessing the normative content of visions of emerging technologies requires to some extent an assessment of the plausibility of these normative visions that emerging technologies will produce desirable worlds.

Following Armin Grunwald, in Chap. 1, I emphasized the importance of engaging in an epistemological analysis of the foundation of an ethical analysis and assessing the quality of the information that is used for such an analysis (Sect. 1.5). As I argued elsewhere, we should not think of "reality checks" as a mere assurance that the information is coming from a knowledgeable informant, such as a scientist (Lucivero et al. 2011). It is indeed better if ethicists and social scientists rely on experts' information concerning emerging technologies rather than on the media, but one of the implications of conducting research on an "emerging" topic is the

2.5 Analyzing Expectations' *Plausibility*: A Proposal

intrinsic uncertainty about its development. Not only is it hard – even for technology developers! – to predict what the final technological outcome will look like, it is even harder to imagine how it will work in society, let alone what effects it will have and whether and how it will be beneficial. Discussing the normative acceptability or desirability of emerging technologies is therefore an arduous task that requires us not only to *weigh* the quality of the information, but also to *reconstruct* pieces of relevant information and inquire about their *plausibility*. In the remainder of this book, I will justify, explain and exemplify the different steps required to address the plausibility question as well as reflecting on how addressing this issue links to the initial question of integrating a normative-sensitive inquiry into TA. My approach is guided by the following reflections.

2.5.2 Breaking Down the Plausibility Question

First of all, in statements about emerging technologies a linear link is often drawn between the technology and the attainment of desirable outcomes:

$$(\text{Emerging Technology} \rightarrow \text{Desirable Consequences})$$

or

$$(T \rightarrow D)$$

Take as an example the visions of genomics described in Sect. 2.1: developments in genomics will make possible desirable medical (and more generally societal) outcomes. I will refer to these projections as "visions of desirability" or D-Visions. As argued above, technologies do much more in interacting with human beings and their morality, and before addressing questions of the desirability of emerging technologies the plausibility of these D-Visions should be assessed. How to do so?

One way to address this question is to break it up into three subquestions:

1. How likely is it that the expected *artifact* will promote the expected values?
2. To what extent are the promised values *desirable for society*?
3. How likely is it that a technology will *instrumentally* bring about a desirable consequence?

The first question focuses on the "T" component of the vision. As explained above, in the materiality of technologies values are "inscribed", roles/responsibilities are distributed among stakeholders, programs of actions are suggested and new interpretations of the world are promoted. To address this question, the expectations of the *future artifact* should be analyzed. Once again, in the case of expectations surrounding genomics, the question proposes to analyze the relevant science and technology and the conditions of its success. Assessing the plausibility in this case means analyzing these expectations and comparing them with the current scientific and technological practice supporting them. It means highlighting challenges and uncertainties and ruling out scenarios that are implausible given the current state of the art. Furthermore, such assessment can have the constructive role of adding new material details and aspects about the technology at stake that can enrich the initial expectations.

The second question focuses on the "D" component of the vision. As the authors of the Vision Assessment approach point out, preferences and ideas of what is desirable are not the same for everyone in society. Different actors will have different perspectives on what is a problem and what is the best solution to address it. Since the new technology is expected to be included in a contextual social practice, with some actors using it in a specific way, the perspectives of these actors are important. Therefore, assessing the plausibility of expectations surrounding emerging technologies also requires gathering different stakeholders' perspectives and expectations pertaining to the use of an emerging technology.[11] Such assessment has the goal of ruling out some scenarios of use as implausible from the perspective of stakeholders. There is also a constructive aspect in this assessment that can add new perspectives and new details about the practice.

These analyses bring us to the last question that pertains the "→"component of the vision: the likelihood that a technology will bring about a desirable consequence. As emerged in the discussion of the literature in the field of philosophy of technology, technologies are not simply instruments to solve a problem. Instead, they change a number of practices, concepts and values. Therefore assessing the plausibility of emerging technologies requires exploring the interaction between technologies and morality. Such assessment has the aim of overcoming a naïve instrumentalist logic according to which a technology will have a desirable impact.

The approach I propose addresses, therefore, these three questions. In order to answer these questions, three interlocked steps are necessary:

- First, the expectations of the artifact should be investigated. As such, the conditions for the technology to work can be explored together with its implications for the context of use.
- Second, the expectations of the context(s) of use should be analyzed. In this way, the social practices together with the values attached to them can be investigated. This helps to explore how the new technologies can be embedded in such practices and systems of values.
- Third, based on these analyses, the plausibility of the expectations that a new technology will realize some values can be assessed. This assessment concerns the three aforementioned questions: what values an expected artifact promotes?; to what extent are the promised values desirable for society?; and how likely is it that a technology will *instrumentally* bring about a desirable consequence?

These are not isolated phases of analysis, but they build on one another. As emerges in the following chapters, what I am proposing is a flexible framework that can be adapted to different contexts, technologies and stages of technology

---

[11] Interestingly, the term "stakeholder" indicates somebody who has something "at stake", some interest, value or (idea of) good to protect.

2.5.3 In Search of Plausibility

In the tradition of ancient rhetoric, a statement is "plausible" or "likely to be true" in a certain context if a specific audience considers it as apparently valid and credible (Aristotle 1954). Etymologically, in the English language, a discourse is "plausible" when it is winning public approval and susceptible to "applause" (Ramirez and Selin 2014). Even without being able to assess the truth of a claim, people in an audience will consider a claim plausible if they can attribute meaning to it and believe that it is convincing (Perelman and Olbrechts-Tyteca1969). As Ramirez and Selin (2014) explain, the aspect of sense-making beyond a proof or evidence of reality and truth is what distinguishes the concept of plausibility from the more quantitative concept of probability.[12] The criterion of plausibility suggests that in gaining information that helps us deal with the uncertainty of the future, "a broader range of future states that are deemed 'occurrable' (even if not very likely)" (Wiek et al. 2013: 144), are taken into account. The aspect of "broadness" highlights that the focus on plausibility is more on the exploration of alternative futures rather than the prediction of probable scenarios.

The aspect of the plausibility of prospective knowledge is related to the criteria of *consistency* and *desirability* (Wiek et al. 2013).[13] A vision is plausible if it is consistent, i.e. it does not present internal inconsistencies, but this is not enough, it also needs to be grounded in reality, in some 'empirical evidence' that justifies the "occurrability" of the vision. Desirability "is neither necessary nor sufficient for plausibility, but it might still be an indirect positive indicator for plausibility" (144). A vision describing a future state that is desirable for a certain audience may sound more plausible to that same audience. It has been noticed that visioning desirable futures are an influential, if not indispensable, stimulus for change (Wright 2010; Wiek and Iwaniec 2013). To some extent a desirable future is also more plausible than a less desirable one, although there seems to be no in depth empirical study on this.

Although less rigorous from the quantitative point of view, plausibility is a better criterion to discuss the future by those who hold a constructivist (as opposed to realist) epistemology that recognizes that a plurality of views, knowledges and values

---

[12] In reconstructing the etymology of the terms, Ramirez and Selin show how the meaning of "plausibility" and "probability" have been confused and hardly distinguishable in some historical phases of the English language.

[13] Wiek and colleagues as well as Selin and Pereira and Ramirez and Selin have discussed the concept of plausibility in the context of foresight tools for decision-making in situations of uncertainty. Often they see plausibility as a criterion for constructing scenarios as tools for deliberative decision-making exercises (see Chap. 7 for a discussion on scenarios).

can coexist and need to interact and confront each other in a meaningful way. When dealing with projections, visions and expectations we cannot rely on facts or observable phenomena, as Cyntia Selin puts it, the notions of evidence and proof are lost and we need to reinvent other modes of what is trustworthy anticipatory knowledge. The notion of plausibility always presents some level of speculation and seems to offer "another mode of relating to the future that acknowledges an intrinsically uncertain and contingent future" (Selin and Pereira 2013: 5).

Alfred Nordmann highlights the difficulties that emerge when trying to use the concept of "plausibility" for future oriented policy recommendations around emerging technologies. According to his conceptual analysis, plausibility can be equated to "serious possibility" as something that "amalgamates the notion of likelihood, of consequence, and of concern" (2013: 126). It is a scenario that is not only technically possible, but also feasible given the specific constraints of a particular world and credible within that given world. The difficulty of judging what is plausible arises when we have to define what the "given world" we are referring to is. Is it the real, actual world that we know from experience or is it some form of imagined "future" world? Additionally, what is the relationship between the actual world and this future world? As Nordmann, critically contends, it is important to clarify that in this case we are in the realm of "plausibility squared" (plausibility^2) where the judgment of possibility has to be made not only with respect to how something is plausible in that given world, but also on how plausible it is that the imagined world is a successor of the present one.[14] Another *caveat* comes from the fact that plausible knowledge maintains its rhetorical dimension insofar as actors try to persuade each other of what matters and what should be taken seriously (Nordmann 2013). Competing plausible visions may disagree with one another and what counts as plausible is "negotiated" among experts and stakeholders in decision-making processes (Selin 2011: 732).

2.5.4 Three Strategies to Appraise Plausible Visions

These reflections explain in what sense talking about assessing visions in terms of their plausibility is more appropriate than referring to "reality checks" or assessments of their technical possibility. Referring to plausibility means to acknowledge the plurality of visions surrounding emerging technologies and to recognize that

---

[14] Nordmann contends that the problem arises when the imagined world is assumed to be a possible likely future of the current world. What is our cognitive access to this future world, and is it qualitatively different from the one we know, asks Nordmann, and what is our ability of determining what will be plausible in this future world? The issue here requires us, of course, to specify how we can distinguish between an epistemically legitimate statement concerning a state of affairs in time in our current world (grammatically using the future tense: "the world population will increase by x number of people during the next 12 months" (133)) and a statement about a "future" which is discontinuous from our known world. Nordmann acknowledges this difference in a footnote, but it could be further explored.

2.5 Analyzing Expectations' *Plausibility*: A Proposal

these visions are not assessed on the basis of their predictive power or realism, but on the basis of the universe of meaning that they bring with them. The credibility of these visions is not reduced to a judgment concerning their technical feasibility but depends on how the world depicted in these visions takes into account a set of societal and cultural aspects. Furthermore, such judgments of plausibility are not univocal and unambiguous. On the contrary, they depend on the value of the person who judges them: whether a scenario is plausible for actors depends on their framework of values and ideas of what is a desirable future. This link between beliefs, knowledges and the moral world is crucial for an approach that aims at assessing plausibility as a condition for discussing desirability questions. Finally, it is important to emphasize that the assessment of plausibility of expectations is not oriented towards a future world that we cannot have access to, but focuses instead on this given world. Analyzing plausibility of expectation should indeed be an attempt to clarify assumptions in visions guiding current decisions. In the three steps of analysis of expectations described above (2.5.2), expectations' plausibility is understood in these terms. In fact, these reflections on plausibility explain the methodological strategies that I use in each one of the three steps of analysis and that tie these three steps together: namely, thickening, zooming in and out, and situating. Let's see.

First, assessing the plausibility of expectations requires *thickening* them. Details are added to their prospective descriptions of how an artifact will operate, how it will be used or how it will have socially desirable impacts. When the general public reads or hears about new technologies through journalists or from the scientists themselves, a big, but quite general picture, is presented before their eyes. The new technology is introduced as a solution in a context where there is a need or a problem. For example, the Human Genome Project offers a solution to satisfy the need for more precise and individualized medicine. This new science and technology is expected to enable the production of innovative diagnostics and therapies targeting specific genes. More generally, early diagnostics enable early and more successful interventions that improve people's health conditions and decrease unsustainable high costs in national healthcare programs. In van Lente's terminology (1992), these macro-level broad scenarios are *nested* with micro-level expectations: they are connected with other expectations that might not be explicitly referred to, but are nevertheless implied in the broader scenarios. For example, the scenario of better diagnostics might entail that genetic information has to be broadly exchangeable and comparable among researchers. The plausibility assessment that I am proposing here turns the philosopher's gaze away from the broad scenarios circulating in the public domain. Instead, the philosopher is invited to evaluate the plausibility of these broader scenarios by focusing on the expectations that are nested within them.

Second, the metaphor of the "mosaic" might help give a better understanding of what I mean when I say that analyzing plausibility entails *zooming in/out*. Imagine the broader scenarios offering the big picture of a mosaic. From a distance the tiles that form the mosaic appear as an appropriate component that contributes to the consistency of the overall image. However, the closer we move towards the picture, the greater number of individual tiles we see and the more details appear. The mosaic tesserae acquire different colors, shapes or textures and the apparent uniformity gives way to inconsistencies and idiosyncrasies. The details of the tiles make the spectator lose the big picture and this is when a step back is beneficial. Now aware of the single tiles, the whole picture can be re-evaluated. Zooming in on the smaller tiles that compose the bigger picture is a recurrent element in my assessment of the plausibility of expectations.

Third, assessing plausibility requires a *situating* strategy. Following insights from the sociology of expectations, I adopt the perspective that expectations of technological futures may vary among different social actors. The plausibility of expectations of a future state of affairs is judged against the background of their beliefs and values. The lay audience might deem some expectations on the feasibility of emerging technologies as plausible, while a group of experts might be more aware of the uncertainties and challenges. For example, within an audience consisting of molecular biologists, geneticists, general practitioners and patients, the expectations of the Human Genome Project appeared more or less plausible depending on their epistemic background. The audience will assess the expectations with respect to their knowledge of physics and electronics, medical practice, self-testing practice or advising patients. Furthermore, these expectations are visions of desirable futures and are carriers of values and ideas of good. Therefore, the judgment of plausibility will also depend on how these different audiences are mobilized by these values.

These beliefs and values change according to the position of the actors. Whether we talk of scientists, policy-makers, doctors, or patients, we can talk of different "epistemic and normative cultures". They will each have a different knowledge, situated and dependent on their background, "expertise", interests and system of values. An exploration of the judgments on these expectations from inside a specific epistemic and normative culture helps in analyzing these expectations. What is plausible for a certain epistemic and normative culture sharing a certain background might not be so for another culture. For this reason, my approach consists of discussing the expectations of stakeholders belonging to different epistemic and normative cultures and drawing attention to controversies and clashes among these expectations. Clashes between visions of future technologies in society within different communities bring up new perspectives. Pointing out alternative perspectives among social actors is important in order to assess the likelihood of some expectations. When a scenario of a techno-socio-moral future is reconsidered from different "situated knowledges" (Haraway 1988) and interests, some aspects of the vision might be re-evaluated as unlikely while others might be added as important. New features enter and enrich the initial picture. Assessing the plausibility of expectations surrounding emerging technologies is akin to fixing a ship while sailing in the open sea; one cannot abandon the ship to look at it from a distance and determine the best means of reconstruction. Such assessment can only occur from within the situated perspectives of social actors in the process of technology development.

These three strategies build up my take on what it means to assess expectations' plausibility and underlie each one of the three steps of analyzes described above (Sect. 2.5.2). The following chapters will explore each of these three steps separately:

they will justify why and exemplify how thickening, zooming in/out, and situating expectations around emerging technologies contribute to an analysis of the expectations' plausibility.

References

Achterhuis, H. 2001. *American philosophy of technology: The empirical turn*. Bloomington: Indiana University Press.
Akrich, M. 1992. The description of technological objects. In *Shaping technology building society: Studies in sociotechnical change*, ed. W. Bijker and J. Law. Cambridge, MA: MIT Press.
Akrich, M., and B. Latour. 1992. A summary of a convenient vocabulary for the semiotics of human and nonhuman assemblies. In *Shaping technology, building society: Studies in sociotechnical change*, ed. W. Bijker and J. Law, 259–264. Cambridge, MA: MIT Press.
Aristotle. 1954. *Rhetoric*. Hg. von W Rhys. Roberts. New York: Cambridge University Press.
Arnaldi, S. 2010. Ordering technology, excluding society: The division of labour and sociotechnical order in images of converging technologies. *International Journal of Nanotechnology* 7(2): 137–154.
Berne, R.W. 2006. *Nanotalk: Conversations with scientists and engineers about ethics, meaning, and belief in the development of nanotechnology*. Mahwah: Lawrence Erlbaum.
Bijker, W.E., T.P. Hughes, and T.J. Pinch. 1987. *The social construction of technological systems: New directions in the sociology & history of technology*. Cambridge, MA: MIT Press.
Borup, M., N. Brown, K. Konrad, and H. Van Lente. 2006. The sociology of expectations in science and technology. *Technology Analysis & Strategic Management* 18(3–4): 285–298.
Brey, P. 2009. Values in technology and disclosive computer ethics. In *The Cambridge handbook of information and computer ethics*, ed. L. Floridi. Cambridge: Cambridge University Press.
Brown, N. 2006. Shifting tenses – from 'regimes of truth' to 'regimes of hope'. Shifting Politics-politics of technology-the times they are a-changin'. *Groningen* April (2006): 21–22. SATSU working papers no 30. https://www.york.ac.uk/media/satsu/documents-papers/Brown-2006-shifting.pdf.
Brown, N., B. Rappert, and A. Webster. 2000. *Contested futures: A sociology of prospective technoscience*. Aldershot: Ashgate.
Burke, W., D. Atkins, M. Gwinn, A. Guttmacher, J. Haddow, J. Lau, G. Palomaki, N. Press, C.S. Richards, L. Wideroff, and G.L. Wiesner. 2002. Genetic test evaluation: Information needs of clinicians, policy makers, and the public. *American Journal of Epidemiology* 156(4): 311–318.
Collins, F.S. 1999. Medical and societal consequences of the human genome project. *New England Journal of Medicine* 341(1): 28–37.
Collins, F.S., et al. 1998. New goals for the U.S. Human genome project: 1998–2003. *Science* 282(5389): 682–689.
Crichton, Michael. 2002. *Prey: Novel*. New York: Harper Collins Publishers.
Dierkes, M., U. Hoffmann, and L. Marz. 1996. *Visions of technology: Social and institutional factors shaping the development of new technologies*. Frankfurt/New York: Campus-Verl.
Epstein, C.J. 2004. Genetic testing: Hope or hype? *Genetics in Medicine* 6(4): 165–172.
Feenberg, A. 1995. *Alternative modernity: The technical turn in philosophy and social theory*. Berkeley: University of California Press.
Fischer, Claude S. 1992. *America calling: A social history of the telephone to 1940*. Berkeley: University of California Press.
Fischer, C.S. 2011. *Still connected: Family and friends in America since 1970*. New York: Russell Sage.

Friedman, B., and Helen Nissenbaum. 1997. Software agents and user autonomy. In *Proceedings of the first international conference on autonomous agents – AGENTS'97*, 466–469. New York: ACM Press.

Friedman, Batya, Peter H. Kahn and Alan Borning. 2003. Value sensitive design: Theory and methods. UW CSE technical report 02-12-01. Seattle: Department of Computer Science and Engineering, University of Washington.

Geels, F.W., and W.A. Smits. 2000. Failed technology futures: Pitfalls and lessons from a historical survey. *Futures* 32(9–10): 867–885.

Goggin, Gerard. 2006. *Cell phone culture: Mobile technology in everyday life*. London/New York: Routledge.

Grin, J. 2000. Technology assessment as a tool for political judgement. In *Vision assessment: Shaping technology in 21st century society V*, ed. J. Grin and A. Grunwald, 9–33. Berlin: Springer.

Grin, J., and A. Grunwald. 2000. *Vision assessment: Shaping technology in 21st century society*. Berlin: Springer.

Guice, J. 1999. Designing the future: The culture of new trends in science and technology. *Research Policy* 28(1): 81–98.

Haraway, D. 1988. Situated knowledges: The science question in feminism and the privilege of partial perspective. *Feminist Studies* 14(3): 575–599.

Hedgecoe, A. 2010. Bioethics and the reinforcement of socio-technical expectations. *Social Studies of Science* 40(2): 163–186.

Hedgecoe, A., and P. Martin. 2003. The drugs don't work: Expectations and the shaping of pharmacogenetics. *Social Studies of Science* 33(3): 327–364.

Holtzman, N.A. 1999. Are genetic tests adequately regulated? *Science* 286(5439): 409.

Holtzman, N.A., and T.M. Marteau. 2000. Will genetics revolutionize medicine? *New England Journal of Medicine* 343(2): 141–144.

Humphreys, L. 2005. Cellphones in public: Social interactions in a wireless era. *New Media & Society* 7(6): 810–833.

Huxley, A. 1969. *Brave new world* [1932]. New York: Harper Perennial.

Ihde, D. 1991. *Instrumental realism the interface between philosophy of science and philosophy of technology*. Bloomington: Indiana University Pres.

Janssens, C.A.J.W., and M.J. Khoury. 2006. Predictive value of testing for multiple genetic variants in multifactorial diseases: Implications for the discourse on ethical, legal and social issues. *Italian Journal of Public Health* 3(3): 35–41.

Konrad, K. 2006. The social dynamics of expectations: The interaction of collective and actor-specific expectations on electronic commerce and interactive television. *Technology Analysis & Strategic Management* 18(3–4): 429–444.

Kroes, P., and A. Meijers. 2001. *The empirical turn in the philosophy of technology*. Amsterdam/New York: JAI.

Latour, B. 1987. *Science in action: How to follow scientists and engineers through society*. Cambridge, MA: Harvard University Press.

Latour, B. 1992. Where are the missing masses? Sociology of a few mundane artefacts. In *Shaping technology, building society: Studies in sociotechnical change*, ed. W. Bijker and J. Law, 225–258. Cambridge, MA: MIT Press.

Latour, B., and Venn, C. 2002. Morality and technology. *Theory, Culture & Society* 19(5–6): 247–260.

Lucivero, F., T. Swierstra, and M. Boenink. 2011. Assessing expectations: Towards a toolbox for an ethics of emerging technologies. *NanoEthics* 5(2): 129–141.

Mike Michael, and Brown, Nik. 2003. A sociology of expectations: Retrospecting prospects and prospecting retrospects. *Technology Analysis & Strategic Management* 15(1): 3–18.

Mitcham, C. 1994. *Thinking through technology: The path between engineering and philosophy*. Chicago: University of Chicago Press.

Mol, A. 2000. What diagnostic devices do: The case of blood sugar measurement. *Theoretical Medicine and Bioethics* 21(1): 9–22.

References

Moreira, T., and P. Palladino. 2005. Between truth and hope: On Parkinson's disease, neurotransplantation and the production of the "self". *History of the Human Sciences* 18(3): 55–82.

Nissenbaum, H. 1998. Values in the design of computer systems. *Computers and Society* 28(1): 38–39.

Nordmann, A. 2013. (Im)Plausibility2. *International Journal of Foresight and Innovation Policy* 9(2-3-4): 125–132.

Nordmann, A., and A. Rip. 2009. Mind the gap revisited. *Nature Nanotechnology* 4(5): 273–274. Nature.

Ofcom. 2012. Communications market report 2012. Available at http://stakeholders.ofcom.org.uk/binaries/research/cmr/cmr12/CMR_UK_2012.pdf

Paul, P. 2011. Don't call me, I won't call you http://www.nytimes.com/2011/03/20/fashion/20Cultural.html?pagewanted=all. Retrieved on 17 Oct 2011.

Perelman, C., and L. Olbrechts-Tyteca. 1969. *The new rhetoric: A treatise on argumentation.* Notre Dame: University of Notre Dame Press.

Quist, J. 2007. *Backcasting for a sustainable future: The impact after 10 years.* Delft: Eburon.

Ramírez, R., and C. Selin. 2014. Plausibility and probability in scenario planning. *Foresight* 16(1): 54–74. Emerald Group Publishing Limited.

Richards, T. 2001. Three views of genetics: The enthusiast, the visionary, and the sceptic. *BMJ* 322(7293): 1016.

Selin, C. 2007. Expectations and the emergence of nanotechnology. *Science, Technology & Human Values* 32(2): 196–220.

Selin, C. 2011. Negotiating plausibility: Intervening in the future of nanotechnology. *Science and Engineering Ethics* 17(4): 723–737.

Selin, C., and A. Guimarães Pereira. 2013. Pursuing plausibility. *International Journal of Foresight and Innovation Policy* 9(2-3-4): 93–109.

Shelley Egan, C. 2011. *Ethics in practice: Responding to an evolving problematic situation of nanotechnology in society.* Enschede: Proefschrift Universiteit Twente.

Swierstra, T.E. 2010. Het huwelijk tussen techniek en moraal. In *Moralicide. Mens, techniek en symbolische orde*, ed. M. Huijer and M. Smits, 17–35. Lemniscaat: Rotterdam.

Swierstra, T. 2013. Nanotechnology and technomoral change. *Etica e Politica* 15(1): 200–219.

Swierstra, T., and A. Rip. 2007. Nano-ethics as NEST-ethics: Patterns of moral argumentation about new and emerging science and technology. *NanoEthics* 1(1): 3–20.

Swierstra, T., R. van Est, and M. Boenink. 2009. Taking care of the symbolic order. How converging technologies challenge our concepts. *NanoEthics* 3(3): 269–280. Springer Netherlands.

van Lente, H. 1993. *Promising technology: The dynamics of expectations in technological developments.* Enschede: Universiteit Twente, Faculteit Wijsbegeerte en Maatschappijwetenschappen.

Van Merkerk, R.O. 2007. *Intervening in emerging nanotechnologies: A CTA of Lab on a chip technology.* Utrecht: Utrecht University, Royal Dutch Geographical Society.

van Merkerk, R., and D. Robinson. 2006. Characterizing the emergence of a technological field: Expectations, agendas and networks in Lab-on-a-chip technologies. *Technology Analysis & Strategic Management* 18(3/4): 411–428.

Van Zwieten, M. 2008. Constructing results in prenatal diagnosis. Professionals anticipating parental decisions. In *Genetics from laboratory to society : Societal learning as an alternative to regulation*, ed. Gerard de Vries and Horstman Klasien. Basingstoke/New York: Palgrave Macmillan.

Verbeek, P. 2005. *What things do: Philosophical reflections on technology, agency, and design.* Pennsylvania State University Press: University Park.

Wiek, Arnim, and David Iwaniec. 2013. Quality criteria for visions and visioning in sustainability science. *Sustainability Science* 9(4): 497–512.

Wiek, Arnim, Lauren Withycombe Keeler, Vanessa Schweizer, and Daniel J. Lang. 2013. Plausibility indications in future scenarios. *International Journal of Foresight and Innovation Policy* 9(2-3-4): 133–147.

Wright, Erik Olin. 2010. *Envisioning real utopias*, vol. 98. London: Verso.

Part II

Chapter 3
The Mechanism in the Pill: From Abstract Images to Detailed Descriptions

> *Caress the details, the divine details. (Nabokov V, Lectures on Literature, 1980)*

Abstract Between 2009 and 2011, TV, national newspapers, magazines, and a children's book presented images of the "Nanopil". A device designed to function as a miniaturized lab within the human body, the Nanopil is expected to test the presence of biomarkers for colorectal cancer and transmit the result to an outside receiver. As the first of three chapters using the "Nanopil" as an example, this one discusses how to analyze expectations about a future artifact. Providing tangible examples from this case study based on in-depth interviews with Nanopil developers and observation within their laboratory, the chapter explores alternative designs, components and conditions of its functioning. The conclusions highlight how such analysis contributes to ruling out implausible scenarios and adds significant detail to the initial public expectations.

Keywords Nanopil • Social construction of artifacts • Laboratory studies • Material design practices

3.1 Visions of Promising Technologies: The Nanopil

> […] It is the idea of the oncologist Prof. Emer. Pinedo, who approached us with the question of whether it would be possible to develop a Lab on a Chip system that can be used for the early diagnostics of colorectal cancer. This disease is the second most common cancer in men and women and is often only picked up at a (too) late stage. (Berg 2009: 38, *author's translation*)

This quote is taken from a speech given by Albert van den Berg, professor of Miniaturized Systems for (Bio)Chemical Analysis at the University of Twente (the Netherlands), on the occasion of the annual ceremony commemorating the founding of the University. Being awarded a few weeks earlier with the prestigious

Spinozaprijs (Spinoza Prize),[1] Prof. van den Berg elaborated on one of the expected outcomes of the research conducted in his laboratory: an innovative device for the early detection of colorectal cancer, the Nanopil.[2]

Between 2009 and 2011, TV, national newspapers, magazines, and a children's book[3] presented images of the Nanopil (NP). This device was expected to perform the task of a miniaturized lab within the human body, testing the presence of biomarkers for colorectal cancer, and to transmit the result of the test to an outside receiver. Although the development of such a complex pill may appear futuristic, Professor van den Berg remarked at the time that there was already a so-called "Camera pill" (PillCam), that takes pictures of the interior of the gastro-intestinal channel and sends them to the outside (2009). Professor Pinedo, the oncologist who conceived of the Nanopil idea, anticipated in 2009 in an interview for a Dutch weekly magazine published by the federation of Dutch professional associations for doctors that in 5–10 years doctors would be able to use the pill in hospital settings (Melchior 2009). In those years, several public debates were organized to discuss the opportunities of and concerns about this device.[4]

Initially circulating in personal communications, funding proposals, or patenting applications, the expectations of the NP travel across environments and circulate within public speeches, science cafés, and the media. The NP is not reality yet, but it already exists as an object of people's praise, desires, fears, investments, decisions, research; in short the NP exists as the object of a broad range of expectations. These expectations directly or indirectly refer to the desirability of this device for early screening of colorectal cancer. A democratic discussion (see Chap. 1) about the desirability of an emerging technology like the NP cannot take these expectations at face value, because they are strategic and context-dependent (see Chap. 2). In order to discuss the desirability of the Nanopil in a meaningful way, we should analyze the plausibility of the visions that this technology will realize certain values. Such analysis aims at ruling out some improbable scenarios and at exploring other scenarios that remain implicit.

The promise is that the Nanopil will be a device enabling early, inexpensive and reliable screening for colorectal cancer. As outlined in Chap. 2, the success of such

---

[1] The Spinoza Prize "is an honorary award for what the winners have achieved in their scientific career". It is awarded by the by the Nederlandse Organisatie voor Wetenschappelijk Onderzoek (Netherlands Organisation for Scientific Research) and it includes an award of 2.5 million euro to spend on research http://www.nwo.nl/nwohome.nsf/pages/NWOA_4XLBF5_Eng?Opendocument&su=no

[2] Since expectations on the "Nanopil" have mainly circulated in a Dutch context, I will keep the Dutch form "Nanopil" rather than the English "Nanopill". Some English reviews of the project circulate on the web, but they are mainly quotations of press releases from the University of Twente.

[3] "Een pil met een lab erin" by Martine Letterie was published by Uitgeverij Zwijsen in 2009 (Letterie 2009).

[4] See for example http://www.sciencecafezeist.nl/index.php/terugblik/3-lezing/15-nanomedicijnen-revolutie-berg.html; Nanotopia: http://www.lux-nijmegen.nl/debat/nieuws/2010/06/10/goed-idee-niet-de-nanopil-gaat-er-komen-dat-zeker and Science Café Enschede: http://www.utwente.nl/tnw/nieuws/archief/2009/2009_02_12/science_cafe_nanotechnologie/

a promise requires that three broad conditions be met. First, the Nanopil will be a functioning artifact: it will detect the presence of markers for colon cancer in vivo, within the intestinal tract, and will communicate the result to the external world. Second, this artifact will be used in the practice of colorectal cancer screening by healthy individuals. Third, the use of this device in the expected social context will have the desirable outcome of decreasing health costs while increasing the possibility of detecting early cases of colorectal cancer, benefitting both potential patients and society at large. In order to assess the plausibility of the expectations about the value of the Nanopil, these three conditions have to be analyzed. The next three chapters will discuss how such an analysis can be performed. In order to assess the plausibility of these visions, the expectations of the artifact and of the context of use have to be assessed. Chapters 3 and 4 will address the question of how this can be done using the case of the Nanopil as an example. Based on these analyses, Chap. 5 addresses the main question of the plausibility of the visions of the desirability of the Nanopil.[5]

3.2 Promises of Emerging Artifacts

> Before a tumor is visible, it is possible to detect some changes in the DNA of the intestine cells. We can develop a pill with special nanowires to which the changing (methylated) DNA in the intestine liquid can bind. The information of the nanowires is sent to a receiver, for example a mobile phone. (Van den Berg 2009: 38, *author's translation*)

This quote, from Professor Albert van den Berg's public lecture for a yearly celebration at the University of Twente (the Netherlands), describes an innovative device in development at the BIOS lab-on-a-chip group[6]: the Nanopil. This minuscule lab that can sample and test intestinal fluid within the intestine has the function of diagnosing colorectal cancer. The words are accompanied by screenshots from a short 1'22" animated video showing the inner functioning of the pill (see Fig. 3.1 in the Appendix).

This video – commissioned by the BIOS group for dissemination purposes – proposes a vision of this future device that is worth analyzing. In this video, a 3D

---

[5] The choice of the Nanopil as a case-study was partially motivated by logistical reasons. First, the research on this device was conducted at the University of Twente and it was initiated at about the same time that I started the research that led to this book in 2008. This allowed me to follow the emergence of a technology concept from the beginning and in discussion with the actors who initiated it. Moreover, the NP received significant public exposure from a very early stage of the project. This public exposure of the technology, and its presentation in terms of its impact on people's lives and medical practice, makes it an ideal candidate for early ethical reflection. Most interestingly, while I was investigating the possibility of working on this case, I observed that this public interest and optimism was counterbalanced by a more cautious attitude on the part of engineers working at the lab-bench with regard to the medical application of their research on NP. This clash between the expectations of insiders and outsiders makes the Nanopil an interesting example of an emerging technology that triggers the imagination of the public and yet requires critical assessment.

[6] See http://www.utwente.nl/ewi/bios/

animated middle-aged man stands against a dark grey background, holding a pill in his right hand. His eyes stare into the void and with an automatic gesture he leans his head back while bringing the pill to his mouth. In the next shot, the pill travels through his esophagus to his stomach and from there to his winding bowel. Then, the image zooms inside the pink cavity of the intestinal tract. Now, the white pill swims in a transparent fluid. Yellow strands floating in the intestinal tract (some of them with red dots attached) enter through a hole inside the pill that is labeled "sampling intake". In the interior of the pill, the yellow strands swim in a white corridor towards a chamber in which many pillars are perfectly aligned. The strands with red dots stick to these pillars. Following a sudden flash, the strands detach from the pillars. They swim to another compartment in which there are railway-like tracks with some strands attached to them. Each entering strand with red dots winds itself around one of the yellow strands on the tracks in a locked embrace. The image zooms out, showing that these tracks are linked to an area in the pill that emits waves. The image returns to the middle-aged man who now has a smart phone in his hand. He looks at it and then holds up the telephone screen to the spectator "Pill status received: result 4.7. Result automatically sent to the physician".[7]

This video offers a scenario of the expected working mechanisms of the Nanopil (NP). From the outside to the inside of the human body, the shooting angle of the camera travels through the intestine and then brings the spectator into the interior of the pill, disclosing the architecture inside the pill. This scenario describes a complex and yet perfectly operating and stable mechanism in which every piece performs a function for the overall purpose: detecting biomarkers for colorectal cancer. The video seems to go inside the black box of the Nanopil, however, as I will argue in the next section, this short film is itself an idealized vision of an unlikely future. If we are pressed by the question whether the Nanopil is a "good" technology, we cannot use this video or the developers' interviews as a starting point to speculate on ethically relevant issues related to this device. Instead, the plausibility of the expectations that this device will operate in certain ways should be put under scrutiny. In order to assess the plausibility of these visions, a critical analysis of the expectations on the artifact is needed. Sections 3.4 and 3.5 will describe how the plausibility of these expectations can be analyzed.

3.3 Rhetoric and Black-Boxes

Presented in science cafes, included in online news or presentations, the video shows a broader audience that the Nanopil is not science-fiction, but a nearby reality. In this sense, we can consider it as a visual rhetorical tool. As a large body of literature in the field of Science and Technology Studies has pointed out the apparently logic, rational and plain narratives of scientific discourses introducing new

---

[7] This video (http://vimeo.com/11547349) was realized by a small Dutch company with expertise in science visualization.

3.3 Rhetoric and Black-Boxes

fields of research (as well as technological fields) is largely rhetorical (see for example, Latour and Woolgar 1979) and always include a certain amount of self-promotion (Guice 1999) that is crucial for technology developers dealing with unfamiliar topics to convince audiences that they are dealing with something real (Brown 2006; Moreira and Palladino 2005). The linear simplicity of the NP system, the causal interaction between its parts and the functionality of its components that appear in the video should not be taken as a neutral description of an existing mechanism. Instead, these features have a strategic function in deleting complexities in order to convey the message to a lay audience that the Nanopil is a feasible device.

This clean and perfectly working image conceals the "dirty" messiness that characterizes the phase of production of technological artifacts, when science and technology are still "in the making", and artifacts are unstable processes of inquiry, inviting heated discussion (Latour 1987). Latour and Woolgar (1979) describe such process of "construction" of scientific facts in the development of what is now viewed as the "central dogma" of molecular biology: the double helix structure of the DNA molecule. In the early 1950s, when Watson and Crick were discussing the configuration of the DNA molecule, the double helix was not a fact, let alone a dogma. Questions, claims, arguments, pictures, enthusiasm or frustration were the only "facts". This process of construction resulted in a scientific "fact": the triple helix became shared knowledge and its process of construction was concealed. As pointed out in Pinch and Bijker's classical example of the bicycle (1987), similar to scientific facts, artifacts are also constructed through social processes; it is only when an agreement is reached that the artifacts stabilize in one specific shape and function. At this end point, they cease to be part of "heated" discussions and become "cold" objects. The process of construction is hidden (or "black-boxed") and its product is presented as an abstract token without a history: an unquestioned and necessary dogma.

At such an early stage of development, a device like the Nanopil is an emerging artifact, still in the process of being constructed. In the video, such a constructive process is "black-boxed" while the expected outcome is presented in its idealized simplicity. The promotional video black-boxes the actual process of construction of the NP. The process of simplification (Callon 1981; Law 1992) has a strategic role in the public space because it shows that the NP is a feasible object. Assessing the plausibility of these expectations about the artifact means looking beyond this simplified rhetoric and disclosing the messy constructive process of artifacts in the making.

This process comprises questions, frustrations and controversies around alternatives. Studies in the history of technology show that, in the emergent phase of technological innovation, different design alternatives coexist. Taking again the example of the bicycle, in its early days, differently designed artifacts co-existed: a model with a front wheel bigger than the back wheel competed with the model with two wheels of the same size that we use today. This second model stabilized as a result of social negotiations and controversies in which different interests were at stake. As Feenberg (1995) remarks, design choices carry a specific vision of the world and system of values. In the case of the bicycle, the model with equally sized wheels was considered to be safer by women and elderly people, while the other model aligned with the interests of young men because it was faster. The fabrication of technical artifacts is a process of "closure" of some controversial and value laden alternative designs within a final design choice. As the case of the bicycle shows, this process of artifact construction unfolds in time and in space: not only do design choices change over time, they are also shaped by controversies and negotiations among actors. This constructive process has been described by reconstructing the history of the artifact and articulating controversies (Pinch and Bijker 1987; Latour 1987). How can this process be analyzed in the case of emerging technologies?

The laboratory offers an excellent site from which to observe: the practices of making science and technology unfold (Knorr-Cetina 1981; Pickering 1995), the controversies and compromises take place (Latour and Woolgar 1979), and the manner in which "accuracy" of a new device is "invented" (Mac Kenzie 1990). Emerging artifacts exist not only in the form of oral or written discourses, like expectations, controversies or patents; in the laboratory, the materiality of emerging artifacts can be observed in experimental setups, biological samples, microscope images.

The laboratory is not only a site to reconstruct the social and material making of the artifact. It is also a site where an emerging artifact can be understood in the epistemic practices of scientists and engineers. In fact, studies of philosophy of science in practice have pointed out that emerging artifacts exist in the laboratory as "models" in which "knowledge of the relevant operations, physical properties, phenomena or properties" are built (Boon and Knuttila 2009). As the authors point out, these models are pragmatic tools with which engineering scientists can intervene in reality. In order to predict the behavior of the mechanism which composes the final product, engineers make use of abstract models with which they "seek to gain understanding of the behavior and properties of various devices and materials" or to produce them. By abstracting from several features of the real device, these idealizations of devices allow engineers to conceptualize a certain phenomenon and to work with it. These pragmatic tools allow engineers to specify a function that a certain device has to fulfill and to highlight functional sub-components. When idealized models are used in the research practice, the relations between the expected overall function of the device and the function of its subcomponents adapt to material and physical conditions. Understanding the idealized model gives a key to access the artifact, as engineers conceptualize it. However, it is also important to go beyond this and look into the materiality of these components.

As Akrich points out, a focus on the "function" of the artifact erroneously suggests that the new technical object is doing the same as the old one in a better (more efficient, precise, sustainable, faster) way. In contrast, by describing some specific aspect of the artifact's material design, attention can be drawn to how an object operates. In particular, she shows how the design process, by defining the components of the artifact, also determines the definitions of actors and relationships among actors. Therefore, in order to focus on what an object does in a social practice, Akrich suggests constructing thick descriptions of the artifact design rather than focusing on the function of the new technology. In order to address the question of the plausibility of expectations for emerging technologies, we have to go beyond the "cold" simplification of the artifact's mechanism offered to a broader audience, as exemplified by the Nanopil video. How to do so?

3.4 A Note on Methods

The Nanopil is still emerging and it does not exist as a fully developed technology. In this sense the NP exists as a vision of an artifact. Not only is this envisioned artifact described in its developers' communications with the broader public, it also appears in research proposals and patent files. The Nanopil exists however also in a more material form: for example, in the components on the laboratory bench, in the differently sized prototypes of the external capsule that are used as reference by the engineers working on it or in the drawings hanging on the laboratory shelves and displaying the different parts of the device. These fictional and material constituencies of the emerging technology constitute a starting point for assessing the plausibility of expectations around it. Following the considerations outlined in the previous section, the reconstruction of the history of the technology at stake offers a crucial tool to highlight the controversies and negotiations around different design choices. The laboratory offers a rich environment to explore technology developers' models and laboratory practices, the different functional components in which they dissect the overall artifact's vision as well as their ideas of how these components contribute to the overall function. The laboratory is also a site where the more material conditions of operation of such components are revealed and the design "choices" are adapted to such material conditions. For these reasons, the laboratory has been the preferred site of the first stage of my fieldwork.

Over a period of 6 months, I visited the BIOS group, at the MESA+ Institute for Nanotechnology at the University of Twente, every 2 weeks. In this time interval, I interviewed all the researchers participating in the project with different roles: the principal investigator, 2 senior researchers, 1 post-doctoral student and 1 graduate student. The BIOS group, included in the faculty of electrical engineering at the University of Twente, "aims at the research and development of Lab-on-a-Chip (LOC) systems", that is miniaturized systems that perform complex biochemical analysis. The researchers I interviewed came from an electrical and bio-engineering background and had experience and interests in nanofluidics and nanosensing. I conducted a first round of semi-structured interviews with each researcher. Although I always began by communicating my interest in understanding more about the Nanopil and its technology, the interviews proceeded in different directions depending on the researcher's interest, attitude and disciplinary background. Recurrent questions concerned the role of the researcher in the project, experienced challenges and reasons for concern and hope. Several interviewees started narrating the story of their involvement in the NP project. One person brought the original project proposal of the NP – and the subsequent interview revolved around the proposal. One senior researcher offered to guide me through the science and technology of the

NP. We initiated a series of bi-weekly meetings. Starting with an in-depth explanation of every scene of the promotional video described above, ever more scientific and technical aspects were addressed. Given my fundamentally humanistic education, I did not have the sufficient disciplinary background to understand all of these aspects; for this reason, I immersed myself in self-study following every meeting. Back at my desk, I listened to recordings of the meetings, transcribed them and collected more information on the scientific and technical aspects that I was unable to understand. When Internet searches or scientific articles could not provide the information I required, I would ask further questions during the next meeting. Some of these meetings took place in the researcher's office and others in the laboratory, where experimental setups were shown and discussed. Since most of the researchers at the BIOS group had an engineering background and they acknowledged their lack of in-house biological expertise for this specific project, I also interviewed a molecular biologist from a different institution who was initially contacted by the BIOS team to provide expertise on biomarkers for colorectal cancer. One year after these meetings, I interviewed the principal investigator at BIOS again about the further developments of the project.[8]

In these interactions, my role alternated between the "interviewer", the "student", or the "philosopher". At the outset, I approached the Nanopil team as respondents of semi-structured interviews, while maintaining sufficient leeway to facilitate elaboration and digression. During the meetings with the senior researcher, I approached the interaction as a student who needs to learn more about the science and technology of the Nanopil. When I became acquainted with the science and I started interviewing other social players,[9] I introduced the inputs of other parties to the conversations, allowing space for critical discussions.

These interactions with technology developers aimed to explore the complexities of an emerging artifact beyond the simplifying strategies of public science. In eliciting their expectations on the artifact in the situated context of the laboratory practice, I focused also on the materiality of their research practice: on the problems that they encountered in their everyday bench research (experimental setup, type of machines, practical challenges). Furthermore, my conversations with the engineering scientists at BIOS aimed to reconstruct the components, functions and aspects that are expected to contribute to the overall functioning of the final artifact.[10] Such

---

[8] The description of my research design demonstrates that the collaboration, availability and patience of researchers at the BIOS group were crucial for my study. I wish to thank them for sharing their knowledge, reflections and time with me.

[9] For an analysis of the input of other social players see Chap. 4.

[10] As one of my respondents pointed out in one of our conversation the distinction into functional component is typical of the "engineering mindset". My analysis of the functional components of the Nanopil was indeed influenced by engineering researchers' way of constructing their project as an assemblage of functional components. In this sense functional component could be considered as "actors' category" as I am not assuming that such components have an ontological or material existence but that this is the way the developers of the NP, think about it and construct it as an object of study (see also Boon and Knuttila (2009) on engineers models). It remains to be seen how researchers with different disciplinary backgrounds would approach the NP project. Although

reconstruction allowed me to "thicken" the description of the artifact that is "blackboxed" in the video. The analysis presented in the following sections describes this type of work. As the last section of this chapter emphasizes, this type of analysis rules out some implausible expectations and enriches the current visions of the Nanopil with new details that are important for the way in which the artifact operates in a social practice.

3.5 The Nanopil: Tales of an Emerging Object

Although all the interviews and meetings were recorded and transcribed, I will not always report their exact transcription, for reasons of readability. At times I will use short narratives that are either summaries of long interviews, or my observations of laboratory practices. These narratives are typographically marked by the *italic font* and indent. They are not meant to be used for close discourse analysis. Instead, following the tradition of "empirical philosophy" (Mol 1998, 2000, 2002), these philosophical narratives have a "heuristic" function; they alert the reader to some observations, interpretations and considerations that guide the plausibility assessment of expectations on emerging technologies.

3.5.1 From an Idea to a Project

The story of a project is a good place to start exploring the complexity of an emerging artifact; in fact, it introduces the temporal dimension and provides a reference point to understand the present stage of research. Nanopil is a special project for researchers working at BIOS. Its concept sprung from the will and mind of an oncologist looking for a technological platform for a specific diagnostic application. Doctor Pinedo was looking for an innovative diagnostic to detect tumors at an early stage and he had the idea of placing this diagnostic in a pill. Pinedo contacted the BIOS group at the MESA+ Institute for Nanotechnologies because of their expertise on lab-on-a-chip systems. These miniaturized platforms perform complex electrochemical analysis of micro-quantities of biological samples and can fit inside a capsule. In this sense, Nanopil can be considered to be a technology "on demand"; the oncologist explains what is needed and the engineers provide the solutions. In practice, however, the roles are not so dichotomously distributed. Since the beginning, in fact, the oncologist had a clear idea of how the device was supposed to work. One member of the BIOS group explains:

---

some differences will emerge in the next chapter regarding the concerns and visions of non-engineering experts concerning the value of the NP and its use, the issue of the difference among "epistemic cultures" (Knorr-Cetina 1999) in shaping emerging technologies has not been the focus of this analysis and would deserve further attention.

One day in December 2007, I received a phone call from the head of the department, he wanted to hear what I thought about this idea of the oncologist. He mentioned this idea of having a diagnostic device that would detect an abnormal state of DNA from inside the body. To enter the body, the diagnostic system had to be small enough to fit into a capsule, like a pill. There are already some laboratories detecting this abnormal state of DNA (that is called hyper-methylation) using some optical methods. However, they need a whole laboratory to do this, it wouldn't fit in a pill. An electrical signal, instead, could be miniaturized in a pill. There are several techniques that can be used, but nanowires seemed a good tool to us. From this overall idea of detecting hypermethylated DNA on a miniaturization platform, we started developing the idea of the Nanopil.

The innovative idea is to detect a specific marker for cancer (hyper-methylated DNA) in vivo and in situ, in other words the pill detects cancer biomarkers inside the human body rather than in the laboratory. These aspects are clearly stated in the application submitted by the oncologist and the medical director of the Dutch Institute for Prevention and Early Diagnostics to the European and US patenting offices. The patent, issued in February 2009, protects the invention of a "device for detecting a medical condition of disease".[11] The inventors want to combine an automated miniature analysis system, or lab-on-a- chip system, for the molecular detection of medical conditions, with a signaling method to notify a subject if a test result is abnormal. They call this combination "in situ lab-on-a-chip signaling (ISLOCS) device". The device performs an analysis in situ by "entering the body" and has a signaling system to communicate the test result to the user (Fig. 3.2 in the Appendix reproduces the image of the device from the patent application). The capsule containing the device can be embodied in different ways, not only by swallowing it, but also by inserting it in the vagina, or in the nose, or ear. Once in the body, the capsule will collect bodily fluid, purify it, detect some marker of a certain disease and transmit the results of the detection to the outside world in various ways (either by releasing a colored dye, or radio frequency or acoustic signals).

In order to be easily introduced in the human body, the device needs to be small, "miniaturized", so that it can be housed in a "capsule" small enough to "enter the body". In the application for patenting, the manner in which the capsule can be embodied is broadly described. In addition, the type of disease that can be detected is not strictly limited to cancer or colorectal cancer. The "Bolus Smart pill System", a device to detect colorectal cancer, is an example of "ideal ISLOCS application" that is presented in the patent document. It can be swallowed, travel through the gastro-intestinal tract, collect intestinal fluid, purify a DNA sample and detect the presence on DNA molecules of an abnormal methylation specific for colorectal cancer. In the event of a positive detection, the integrated electronic system commands the release of a blue pre-stored dye that is pumped into the intestine, coloring the stool that can be observed after defecation.

---

[11] "Device for detecting a medical condition or disease" by Pinedo H. M. et al. United States patent application, February 24, 2011, US Patent and Trademark Office, http://appft1.uspto.gov/netacgi/nph-Parser?Sect1=PTO1&Sect2=HITOFF&d=PG01&p=1&u=/netahtml/PTO/srchnum.html&r=1&f=G&l=50&s1=20110046458.PGNR

3.5 The Nanopil: Tales of an Emerging Object 75

The Nanopil is a device that has the function of detecting diseases in the human body. The novelty of this invention lies in the detection of markers for diseases inside the body by an automated device, which communicates with the external world. Colorectal cancer is an ideal disease to be diagnosed with this technology because the capsule can easily be embodied by ingestion. Travelling through the digestive tract, the device can detect cancer biomarkers in it. The application of the "nanopil" idea for the detection of colorectal cancer is possible because the device is small enough to be ingested, and that the digestive tract is continuously unfolding from the mouth to the anus. Thus, the "Bolus smart pill system" is made physically feasible due to a physical characteristic of the device, together with a physical characteristic of human bodies.[12]

This story of the initial idea of the NP is interesting because it articulates its "core" concept. This story provides a narrative background for the researchers to justify or explain their research. As the following shows, these ideas of what the device should do are implemented and translated in research practice.

3.5.2 *An Idealized System and its Building-Blocks*

When I first entered the BIOS lab, the senior researcher and a graduate student were setting up a complex machine sitting on the lab bench. They wanted to optimize the signal detection of the nanowires and compare it with other detection systems. In their interactions, the two engineers did not mention the Nanopil or its use. Afterwards, they explained that the Nanopil as such is mentioned in the lab only when doctor Pinedo goes to visit them: or when people, like me, ask explicitly about the pill. During their daily activities, conferences, and in posters or presentations, they talk about the nanowires and not about the Nanopil *per se*.

The Nanopil as final device was the subject of communications to a broader audience, but it appears not to be a "plausible subject" in lab-talks or scientific conferences. In these research contexts, the researchers' activity and discourse focus on NP components and sub-components. The artifact-in-the-making is discussed in a variety of ways in which the initial abstract Nanopil-object and activity is replaced by other objects and activities. The BIOS researchers' attention is directed towards the suitable temperature for the DNA molecule for proper hybridization; or to investigating the appropriate salt concentration; or to inquiring into the working mechanism of the pumping system.

---

[12] It is interesting to note that, in the context of the research proposal, an additional reason is put forward to support the application of the pill for the detection of colorectal cancer. This second reason concerns the high benefits deriving from the early detection of this type of cancer, since it is a relatively slow-growing cancer and the surgical intervention for the removal is relatively easy. Therefore, claims about the technical feasibility of the pill are intertwined with claims about its social desirability. I will come back to this point in the next chapter.

When I first asked about the technology of the Nanopil, a senior researcher showed me the dissemination video. Afterwards he explained that, in order to understand this complex system, it had to be "broken up" into "building-blocks". During that meeting and subsequent meetings, drawings and further elaborations enabled me to understand that these functional chambers or compartments within the pills allow the following activities:

1) **Sample collection and purification:** When the capsule reaches the bowel through the digestive tract, the intestinal fluid is sampled through a hole at one extremity of the pill. A pump brings the intestinal fluid into the pill. The pill is then directed into a chamber with glass pillars. Only DNA molecules with hypermethylated regions groups on some distinct areas stick to the pillars. The reset is washed out of the pill via another pumping mechanism. Then, the hypermethylated DNA is detached by the glass pillars and transported to the detection chamber.
2) **Detection of marker:** The hypermethylated DNA is conveyed to a chamber where complementary single stranded DNA molecules are attached to the silicon nanowire sensors. These DNA molecules present a complementary base sequence to the gene associated to the tumor that has to be detected. Thee nanowires are similar to electrical wires, but at the nano-scale. A certain voltage is applied to the nanowire. If the hypermethylated DNA retrieved from the intestinal fluid presents the gene mutation, it hybridizes with the complementary DNA molecule on the nanowire, and it produces a change in the electrical current in the silicon nanowire sensor.
3) **Result notification:** The change in current in the nanowire sensors is sent via electrical circuit to a chip (similar to a computer chip) which controls the whole mechanism. It sends electrical signals to a belt that the user has to wear in order to amplify the signal that is then sent to an external device (like a mobile phone), which translates it into numeric data on a read-out.
4) **Electronic interfacing**: or the chip, that provides the "brain" of the system
5) **Electrical power supply**: or the battery, which constitutes the "engine" of the device"

The images in the Appendix (Figs. 3.2 and 3.3) illustrate the different chambers that will compose the Nanopil. These boxes are viewed by researchers as signposts for the main components of the device. These signposts are "abstractions" in the sense that the detail and the exact working mechanism are still unknown and need to be explored. Therefore, after the overall function of the NP has been defined, engineers' expectations focus on the specificity of the single functional components. In this sense, both this image and the video of the NP do not have to be taken as representing the exact way the pill will work. On the contrary, they have an "intervening and constructive character" (Boon and Knuuttila 2009): they are pragmatic tools.

This distinction in functional components constitutes a pragmatic tool with which to manage such a complex and big project. For example, each one of these functional steps has become a research project with specific sub-goals and steps. By highlighting the most important functional components in the project, researchers set priorities. They may decide to focus on those parts of the project for which the feasibility needs to be demonstrated in order to prove that the Nanopil is feasible *in principle*. As emerged during a conversation with the senior researcher, the sensing function is a prior goal for proving that the NP works in principle. Such a priority is determined by researchers, based on the fact that the defining characteristic of the device is its reliability, providing a marginal rate of false positives and, most importantly, false negatives. Such reliability can only be demonstrated by the appropriate functioning of the component detecting the cancer's marker. Since researchers establish that the main function of the NP is to perform highly selective and specific detection of a molecule, the proof of principle for the NP consists in providing evidence that the marker detection with nanowires is possible, reliable and sensitive.

The functional diagram of the Nanopil is only an idealized abstraction that provides a working tool for engineers. This abstraction of the "important" features of the real device has the role of making things easier to handle in a research setting in which engineers can focus on one aspect at a time.[13] In daily research work, the final expected artifact, the Nanopil, is only an end goal, while the researchers' focus is on single sub-problems. While addressing these questions on specific problems, through literature reviews, trial and error at the lab bench, frustrations and successes, it emerges that what was considered feasible in the first functional diagram becomes implausible.

> When Doctor Pinedo pays his monthly visit to the laboratory, in December 2011, a researcher explains that the original idea of having molecules stand vertically on the nanowires proved difficult to implement. In order to have optimal conditions for these molecules to hybridize they have to lie horizontally on the nanowires.

The positioning of the nanowires can be adapted in order to ensure that the main function of the NP is performed in the optimal way. The initial concept of the NP and the idea of the overall function of the device is not directly discussed by the BIOS researchers, however it informs their research practice. This aspect is even more evident when the design choices of NP developers are compared with the perspectives of the molecular biologists who had been co-opted for the realization of this device.

> While the BIOS group maintains acknowledged expertise in the miniaturization field, the researchers, predominantly engineering scientists, have to rely on the expertise of molecular biologists with regard to the choice of the relevant markers and their statistical relevance for screening purposes. I went to visit a researcher working on tumor biomarkers who was in contact with the Nanopill developers; at the beginning of the project she provided them with information on cancer biomarkers. She explains the complexity of the tumor formation to me: "we will always need to detect a panel of biomarkers, one is not sufficient". Back to the laboratory, the researchers explain to me that they are working on the detection of one biomarker: the DNA hyper-methylation. What if molecular biologists think that a single biomarker is not enough, but you need a panel that comprises several of them? "Well, multiple biomarkers would increase the sensitivity and specificity, but then you would lose in terms of miniaturization and velocity of the system. If the idea is to have a system that everyone can have, then we have to go for an innovative system like the Nanopil. Our goal right now is to assess the efficacy of the detecting system - the marker that we use for this is not our main focus"

The molecular biologist focuses on the complex interplay of different mechanisms and aspects that need to be taken into account in order to ensure the sensitivity and specificity of the test. Instead, the researchers at BIOS concentrate their

---

[13] In the philosophy of science, these abstractions are conceptualized by the notion of ceteris paribus clauses: "the study of some group of tendencies is isolated by the assumption that other things being equal" (Alfred Marshall) "the more the issue is narrowed and the more it can be handled: but also the less closely it corresponds to real life".

activity on the assembly of a miniaturized and fast device for in vivo analysis. It seems that the divergence of opinion between the two researchers goes beyond a scientific question of which biomarker is more predictive; instead, the difference in opinion concerns the overall goals in the research on diagnostics: the portability of the device versus the sensitivity and specificity of the result.

3.5.3 The Functional Components and Their Material Conditions

Researchers at BIOS are focused *inter alia* on their experimental setups, handling wires, locating glass supports, making calculations, designing circuits, reading scientific articles, and discussing their ideas. They focus their attention on specific sub-components of a bigger system. However, even when working on concentrations, binding events and micro-pumps, the link with the final product is still visible to the attentive eye.

> A researcher is in the lab: beside his work bench there is an A4 sheet with three drawings of capsules of different sizes. He is in charge of the pumping component.
>
> "This is the real challenge of the project," he explains to me "we know that we will manage to do something with nanowires, but having a pump that is so small and so powerful, is difficult. We need the pump to be fast enough to sample the required amount of intestinal fluid in a relatively short time. We managed to make it work in 15 hours, but this is too long. The pill has to collect a relevant amount of samples and detect it in few hours, when the sampling conditions are good".

The optimal pumping time is defined by the digestion time of the user. Despite the focus of BIOS researchers on single components and sub-functions, as explained above, the optimal criteria of the mechanism are dictated by the overall function of the artifact of which these components are part. However, the opposite is also true: physical conditions might shape the performance of the Nanopil in the context of use. As explained by the principal investigator:

> In order to be able to sample the intestinal fluid, the viscosity of the fluid should be low enough so that the pill doesn't get stuck into solid feces. The stool has to be softened. A laxative would do the trick. It is the same procedure that they have before a colonoscopy.

The pill needs to be able to swim through a fluid substance. This is not always the case in the human intestine. In order for this condition to be met, the user has to swallow a laxative. On the one hand, the overall criteria for the optimal performance of each component are established in relation to the final overarching function of screening for colorectal cancer. On the other hand, the decisions at the lab bench – while shaping the performance of (a) single sub-(sub-sub-)component – also define the performance of the NP. Even if the models used by engineers at BIOS are abstractions of real, concrete situations, they set a standard for the behavior of the artifact in normal conditions. For example, they focus on physical/technical conditions for the NP to sample intestinal fluid at low viscosity. Such a condition is

3.5 The Nanopil: Tales of an Emerging Object

satisfied by the use of a laxative by the user that softens the stool in the intestine. These models set some conditions of normality that should be respected in order for the mechanism to work as expected. Some of these conditions are likely to be embedded in the design of the artifact and determine the use of the final artifact, even though they do not clearly appear in the broader scenarios on the NP.

Similar to this example, other technical "details" related to the feasibility of the artifact require modification of the initial expectations concerning the Nanopil.

> Two researchers show me their experimental setup. They want to measure the electrical signal of the nanowires. At the moment, they don't have any DNA probe on the nanowires; it will take few days for the probes to stabilize, before they can measure the electrical signal. "But in the pill, the probe will be already on the nanowire, it will be sold that way, they will have a protective layer that has to be removed before you do the detection", they point out. So can the probe stay on the nanowires for a long time? "No, these molecules that are on the surface have a relatively short lifetime. But this is not a problem. Let's say, for example, that you have the pills on the shelf and then you have sent the probes for different types of cancer. Then somebody orders 10 pills and before you ship the pill you can put different probes on it. This way, you don't have to worry too much about the life time of the probe, because you know that it will be used in 2 weeks or so, so you can put the probes as you need them".

The short life span of the DNA probes on the nanowires requires them to be added to the pill just before use. This technical condition implies that the pill cannot be stored for long periods in a pharmacy. One way of addressing this requirement is to assemble the Nanopil on demand, just before the pill has to be used. Once more, the technical conditions for the pill to work redefine the scenario of its use in a social context, adding relevant aspects that are excluded in the original expectations.

> The set up equipment that is used to measure the electrical signal is cumbersome, it occupies half of the space on the lab bench, thick wires are all around it. How can such a big machine be miniaturized in a chip? "The system to measure the electrical current can be made very small with a computer chip", they explain, "the reason why we need to have these external special cables is to protect the signal from external influences. On a chip it is easier; everything is located in a small place within a few micro-meters. The reason why we do it this way now, with such big machinery, is that every time you have a special chip made, it's around…anywhere from 10.000 to 15.000 euro. This is the price if you need one or two chips made. If you are mass-producing you can have them for 2 euro. But for only one chip, it is very expensive."

The computer chip, which constitutes the main platform of such a miniaturized system (lab-on-a-chip), is too expensive to be manufactured in small amounts. For the final pill to be cheap, chip production should be high.

Finally, at such an early stage alternative subprojects might run in parallel in order to develop the most efficient mechanism to perform a certain function. Take, for instance, the mechanism of notification of the test result: two alternative systems are currently explored by researchers at BIOS. Doctor Pinedo's original idea (articulated in the patent) was to have a blue dye stored in the pill that would be released in the event of biomarker detection and be visible in the stool of the user. This mechanism has been explored by engineers at BIOS. However, they propose an alternative mechanism for the notification of the test result: a radio-signaling system that sends the signal to an outside device, like a computer or a mobile phone. According to the Nanopil developers, this system is preferable because it would allow the integration of a technology that is already in place for other devices (like the PillCam). The blue bolus system, by contrast, requires the design and miniaturization of another pumping system, a task which is seen as a burden for Nanopil developers.

In the patenting application, the blue dye system is emphasized whereas the video discussed above demonstrates the radio-signaling system. While broadly targeted discourses on emerging technologies tend to provide a unified vision of the technically possible artifact, an analysis of researchers' activities and decisions discloses the different designs. As in the case of the radio-signaling, one design might be preferred over another because it seems more feasible, however, as the next chapters will show it is not only a matter of feasibility. In fact, these two designs acquire a different meaning in a different context of use.

3.6 From the Lab "Details" Back to the Big Picture

Analyzing the plausibility of visions of emerging technologies requires an analysis of the expected object of these expectations: the material artifact. The descriptions of the expected artifact aimed at a general public are often simplified abstractions of these objects. Whereas these public expectations describe the future artifact as an accomplished cold device, the laboratory offers interesting access to expectations, in which the history and making of the artifact is still visible. Within this space, researchers' choices, uncertainties, challenges, controversies and doubts, concealed in public expectations, become visible. Furthermore, in research practice, alternative design choices co-exist and are discussed and some design aspects that were too complicated to be explained to a broader audience emerge.

Thus, in addressing the case of the Nanopil, my strategy was to dive into the midst of the techno-scientific[14] practice in which the techno-scientists are "in action". I presented some considerations collected during my study of the pill-in-the-making within the laboratory. My approach to the analysis of the emerging artifact was based on the two strategies introduced in the previous chapter: situating and thickening. I elicited the expectations in the situated practice of the artifact making, where expectations are embodied in experimental setups and development tasks. New aspects of the expected artifact, collected during the situating exercise, were added to the original expectations, while others were removed, resulting in "thickened" descriptions of the artifact-to-be. This type of analysis

---

[14] "Technoscience" is a concept used within actor-network theory (see Latour 1987) to blur the difference between scientists and engineers. According to this theory, science and technology involve the same process of creating larger and stronger networks of human and non-human actors.

highlights uncertainties, controversies and additional challenges that might constrain or redirect the development of the technology. While some scenarios of use can be ruled out as implausible given the current state of the art, others can be enriched with details of how the artifact will work. In any case, the temporal dimension of the technology development process has to be taken into account. When looking at this process, it is evident that there are many variables at stake. The Nanopil might turn out to have different components and functions from the ones expected now. Furthermore, it might transpire that the device itself is unfeasible. Should we just avoid talking about the Nanopil as the end product of research? I argue that we should not; as I explained in the previous section, the end product plays a role in making choices in designing physical components or in development practice.

Reconstruction of the history of the project proposal and patenting application allows the main concept and overall function embodied in the Nanopil to be highlighted. These general ideas on the function of the end product play a role in defining its functional components. In their daily research, technology developers deal with the problems, challenges and uncertainties involved in single sub-projects. In doing this, they are influenced by general ideas regarding the end-function of the entire artifact. This investigation of the expectations on an emergent technology in the laboratory provides a different angle to the one offered by public promises. In this way, initial hype is reduced to more humble expectations. Furthermore, when exploring the material aspects and conditions for the artifact to work, some aspects can be elucidated that transcend techno-scientific considerations. For example, in order for the NP to work the user has to ingest a high amount of laxative. The conditions for the adequate performance of the artifact also set a standard of use, determining how the pill should be used. An investigation of these conditions can contribute to re-adjusting the macro-expectations and public discourses related to it. Take as an example the observation that DNA strands located on the nanowires have a short lifespan. This technical condition has relevant social implications for the retail of the Nanopil. Greater attention to "technical details" rebalances the discourse on the expected or desirable use of the Nanopil on less speculative grounds: this overcomes heated debates on the desirability of an improbable scenario in which the Nanopil is available on the shelf at the grocery store like a pregnancy test.

Finally, the stories from the lab floor suggest that the description of the artifact is not given once and for all. On the contrary, it develops over time and across actors. Whereas broader promises tend to depict the expected artifact as a "cold" established product, in the laboratory the expected artifact takes the form of an unsettled "warm" process. Several artifact designs compete or co-exist. As the following chapter will show, these artifact designs entail a different context of use. An assessment of the plausibility of these expectations can contribute to this research practice by making these scenarios of the Nanopil more robust. This will be further explained in the next chapter, which delves into an analysis of the expected context of use of the

Nanopil. The Nanopil is expected to be used within a certain clinical practice. But what are the working conditions of this practice? How do actors in this practice consider the plausibility of the expectations on the use of the Nanopil?

Appendix

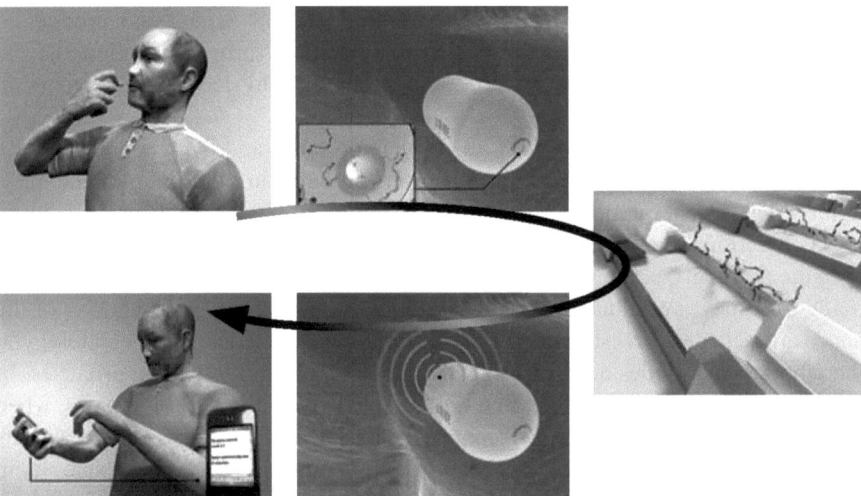

Fig. 3.1 Stills from the promotional video used in Berg 2009: 39

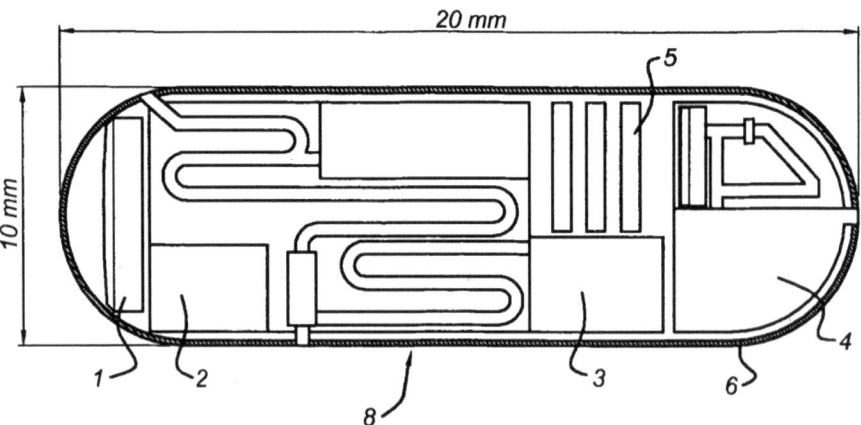

Fig. 3.2 Reproduction of the different chambers in the "in situ lab-on-a-chip signaling device" from the patenting application, available at http://www.wipo.int/patentscope/search/en/WO2009104967

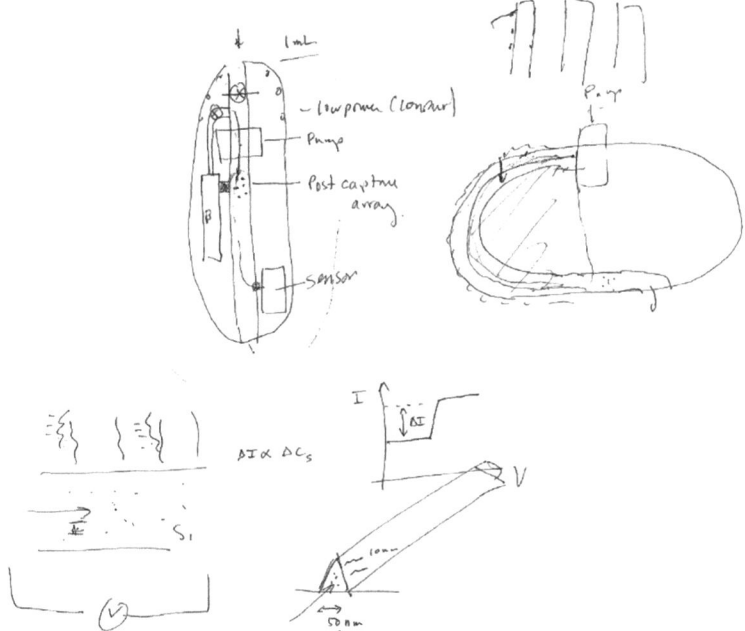

Fig. 3.3 Model of the Nanopil drawn by a researcher at the BIOS group during an interview

References

Boon, M., and T. Knuuttila. 2009. Models as epistemic tools in engineering sciences: A pragmatic approach. In *Handbook of the philosophy of technological sciences*, ed. A. Meijers, 687–719. Amsterdam: Elsevier Science.

Brown, N. 2006. Shifting tenses–from "regimes of truth" to "regimes of hope." 'Shifting politics-politics of technology-the times they are a-changin'. Groningen, April (2006): 21–22. SATSU Working Papers no 30 https://www.york.ac.uk/media/satsu/documents-papers/Brown-2006-shifting.pdf.

Callon, Michel. 1981. Pour Une Sociologie Des Controverses Technologiques. *Fundamenta Scientiae* 2(3/4): 381–399. Mines ParisTech París.

Feenberg, A. 1995. *Alternative modernity: The technical turn in philosophy and social theory*. Berkeley: University of California Press.

Guice, J. 1999. Designing the future: The culture of new trends in science and technology. *Research Policy* 28(1): 81–98.

Knorr-Cetina, K. 1981. *The manufacture of knowledge: An essay on the constructivist and contextual nature of science*. Oxford: Pergamon.

Knorr-Cetina, K. 1999. *Epistemic cultures: How the sciences make knowledge*. Cambridge, MA: Harvard University Press.

Latour, B. 1987. *Science in action: How to follow scientists and engineers through society*. Cambridge, MA: Harvard University Press.

Latour, B., and S. Woolgar. 1979. *The social construction of scientific facts*. Beverly Hills: Sage.

Law, John. 1992. Notes on the theory of the actor-network: Ordering, strategy, and heterogeneity. *Systems Practice* 5(4): 379–393.
Letterie, M. 2009. *Een pil met een lab erin*. Tilburg: Uitgeverij Zwijsen.
Mackenzie, D.A. 1990. *Inventing accuracy: A historical sociology of nuclear missile guidance*. Cambridge, MA: MIT Press.
Melchior, M. 2009. Doctoren met nanotechnologie. *Medisch Contact* 64(nr 49): 2032–2035.
Mol, A. 1998. Missing links, making links: The performance of some artheroscleroses. In *Differences in medicine: Unravelling practices, techniques and bodies*, ed. A. Mol and M. Berg, 141–163. Durham: Duke University Press.
Mol, A. 2000. What diagnostic devices do: The case of blood sugar measurement. *Theoretical Medicine and Bioethics* 21(1): 9–22.
Mol, A. 2002. *The body multiple: Ontology in medical practice*. Durham: Duke University Press.
Moreira, T., and P. Palladino. 2005. Between truth and hope: On Parkinson's disease, neurotransplantation and the production of the "self". *History of the Human Sciences* 18(3): 55–82.
Pinch, T.J., and W.E. Bijker. 1987. The social construction of facts and artifacts: Or how the sociology of science and the sociology of technology might benefit each other. In *The social construction of technological systems: New directions in the sociology & history of technology*, ed. Wiebe E. Bijker, Thomas Parke Hughe, and Trevor J. Pinch. Cambridge, MA: MIT Press.
Pickering, Andrew. 1995. *The mangle of practice: Time, agency, and science*. Chicago and London: University of Chicago Press.
van den Berg, A. 2009. De kunst van het kleine, in Brinskma and van den Berg, De kunst van de wetenschap, Redevoeringen 48ste Dies Natalis, Universiteit Twente.

Chapter 4
The Doctor in the Pill: From "Technical" Details to Social Practices

> *Right from the start, technical, scientific, social, economic, or political considerations have been inextricably bound up into an organic whole. (Callon 1987: 84)*

Abstract This chapter addresses the questions pertaining to the analysis of expectations around the use of emerging technologies. As in the previous chapter, the Nanopil – an emerging artifact for colorectal cancer screening – offers a case study to address the central question of this chapter. The concept of a "fictive script" is introduced as a useful analytic tool to investigate scenarios of use and application of an emerging technology. The fictive scripts conceived by designers are compared both with the script inscribed in the technological object, and with alternative "worlds" as conceived by other actors. Following a short explanation of the research design used for investigating these questions in the case of the Nanopil, the findings of this analysis are presented and critically discussed.

Keywords Actor-worlds • Fictive scripts • User studies

4.1 Expectations of Artifacts in Use

Expectations concerning an emerging device not only describe how its technological apparatus will be able to perform a specific function, but they also frame the device within a broader social context. For example, in the case of the Nanopil – introduced in the previous chapter – the description of how a computer chip inside a small capsule is able to detect a biomarker for colorectal cancer is presented together with the social need that the device is expected to address.

> Colonic cancer is one of the most common cancers in people over the age of 50. The Dutch Health Board has already advised endoscopic or colonoscopic screening for people in this age group. But this is a painful and uncomfortable experience. Moreover, it presents a logistical nightmare and nothing is found in 95 % of cases. What we need is a simple first-line

test. The only alternative at present is a faeces test, but eventually, a nano-pill will provide a much more patient-friendly alternative.[1]

The new device is expected to be employed within a broader social setting: a context of use is introduced, a medical practice is described and the role of institutions or specific social groups/actors is proposed.

> Albert van den Berg [professor of Miniaturized Systems for (Bio)Chemical Analysis at the University of Twente, The Netherlands] clearly imagines the use of this pill: a population screening program in which, for example, every three years all people aged over 50 are invited to swallow a nanopill. In the event of a positive result, a colon examination will follow. "I heard from the medical specialists that in this way the number of examinations, surgeries and deaths from colon cancer will decrease sharply. After all, the sooner something is found, the easier it is to treat the tumor." (Melchior 2009)

The Nanopil is introduced as a screening device, its context of use is sketched and the plausibility of the overall vision is supported by the opinion of other professionals, medical specialists for instance. The expectations that the Nanopil will realize a desirable world are based on expectations of how this device will operate in a certain social context and how it will be employed by users. But how can we assess the plausibility[2] of these expectations of future use?

4.2 (Fictive) Scripts and Actor-Worlds

The concept of "script" (Akrich 1992; Latour 1992) was introduced in Chap. 1; it explains how the design of artifacts not only "embodies" some assumptions about use and users but also "prescribes" how they should be used. I employed this concept as a means of underlining the importance of looking at the design choices and assumptions of technology developers when assessing the plausibility of visions of emerging technologies. Depending on the choices made in designing the artifact and on the assumptions inscribed in it, the final product will "prescribe" different behaviors and "correct" uses. The previous chapter has shown how some of these assumptions of use can be identified in the current research on the Nanopil. There is yet another aspect of Akrich/Latour's script analysis that is relevant when assessing expectations: the assumptions of designers about users and uses do not always correspond to the "real" users/uses.

Akrich emphasizes that "the world inscribed in the object" might differ from "the world described by its displacement" (1992): the user's environment that is specified and configured by the technical object may not match the actual object-user environment. In order to compare the projected use/user with the real one, Akrich travels to countries (specifically, less developed countries) where the new technical object (a photovoltaic generator) is employed and interviews users.

---

[1] A Pill with a Lab Inside http://www.azonano.com/news.asp?newsID=15039 Posted December 8th, 2009.
[2] See Chap. 2 for a discussion about "plausibility".

Specifically, she asks the users how the practice around electric generators functioned before and after the photovoltaic generator was in place. In fact, the "program" inscribed in an artifact does not determine the behavior of the user in a linear way. Mechanisms of adjustment between the real and imagined user may occur or the "program of action" inscribed in the object may conflict with an "antiprogram" of some users.[3] Users can define roles by themselves or can resist the design. Take the example of the program of action inscribed in the design of new cars, whereby the alarm starts beeping when the car moves while the seatbelt is unbuckled (see Latour 1992). Users can resist this program by, for example, jamming something in the seatbelt buckle, or by buckling the seatbelt but passing it through the back of the driver's seat rather than from the front.[4]

These considerations emphasize that not only the artifact and its design but also users' behaviors shape social practices in which new technologies are introduced. According to researchers in the field of *Users Studies* – which builds on history of technology, innovation studies, feminist studies and social constructivist theories – the idea that users merely adapt to new technologies is a highly problematic simplification. Technologies are re-inscribed by consumers in logics that are different from the ones they were conceived by designers. This literature blurs the distinction between production and consumption (Oudshoorn and Pinch 2003) emphasizing that potential users (and social groups) play a role in the design of a certain technology (like the case of the bicycle (Pinch and Bijker 1987) or that of the scientific instrument industry (Hippel 1988)), and contribute to major innovations of the new products and instruments they use. Besides the phase of technology development, users also play an important role when the product is on the market. Furthermore, when the technology is stabilized, the user will embed the technology in a certain practice, creating new practices and processes of appropriation sometimes unexpected by the designers (Silverstone et al. 1992).

While the concepts of script, script analysis and user focus on existing objects, whose actual use can be described, de Laat (1996, 2000) notes that this concept can also be referred to expected objects. In the case of emerging technologies, designers assume that users and other actors will relate in a certain way. As for scripts of material objects, there might be some gaps between the designers' assumptions and the present situation in expectations on emerging technologies. However, since the objects of expectations are still under construction, these assumptions cannot be de-scribed by analyzing material design choices. Emerging objects often exist as *texts* in funding proposals, public oral communications, interviews or patents; for this reason, their "description" can only have a "fictive" character. In developing an analysis of "scripts of the future" or "fictive scripts", the author refers to Michel Callon's concept of "actor-world".

---

[3] In the script-vocabulary, if a user refuses to do what is "prescribed" by the artifact, s/he refuses to "subscribe" to the artifact's prescription.

[4] I have personally witnessed drivers in the city of Naples (Italy) adopting this type of "anti-program".

Callon introduces the concept of "actor-world" as a means of explaining the fact that techno-scientists' expectations on new and emerging technologies are constructions of entire worlds rather than descriptions of a technology. This is illustrated through the example of the electric car (1986, 1987). In 1973, the engineers at the French electricity production/distribution company EDF (Electricité de France) wrote a plan for the introduction of an electric car. In a society dominated by the traditional automobile, EDF engineers described not only the technical characteristics of the electric vehicle (VEL), but also the social universe in which it would function, with the groups composing it and its trends of change. Different types of "entities" were included in this plan.[5] Indeed, the components or "ingredients" of the VEL were non-human entities, like the catalysts to be used in the batteries, and human-entities, like the automobile company Renault that was expected to invest in the production and distribution of vehicles or the emerging environmental groups, that were expected to support the VEL project as a symbol of a post-industrial society.

In an actor-world, several actors, human and non-human, appear together in "heterogeneous associations".[6] In Callon's terms, the techno-scientists *juxtapose* different entities in the context of their visions by *simplifying* them. "Simplification" implies, for example, that entities like towns are reduced by EDF engineers to some limited aspects such as city councils in charge of developing non-polluting transport systems. Such simplification is necessary as a means of accounting for an "infinitely complex world" (Callon 1986: 29). Thus, by reducing entities to a few properties that make them compatible with other inter-related entities, the "actor-world" appears consistent and coherent.

In an "actor-world", a particular actor (say a techno-scientist, an activist group or a governmental body) imbues other entities with a certain identity, task, or concern; in addition, some roles are attributed to them.[7] This attribution of identity and role is an expression of the interest of the actor who is designing a certain world. In the case of the electric car, the roles and identities attributed to social groups, Renault engineers, or catalysts to be used in the batteries were an expression of the interest of the EDF engineers. These constructed worlds were dependent on the knowledge and interest of a particular actor and can differ from the ones that other actors, say the engineers at Renault, would depict.

---

[5] The socio-constructivist tradition emphasizes that technological innovations are not simply physical artifacts, but more complex "technological systems", made of different components, including also legislative artifacts (regulations), social or political organizations and institutions (Hughes 1983, 1987). In this sense, regulations, social groups and institutions are not part of the social environment – or the context – of a technological system, but they are part of the system itself. An analysis of these systems should reject the distinction between the inside (technology) and the outside (society) (Bijker et al. 1987). The concept of "actor-networks" (Callon 1981) emphasizes that these systems are composed of both human and non-human elements and that a distinction among them is not productive.

[6] Callon describes "these heterogeneous associations and the mechanisms of their transformation or consolidation" using the notion of "actor -networks" (1987: 93).

[7] In Callon's words, actors are "enrolled" (Callon 1986).

4.2 (Fictive) Scripts and Actor-Worlds

The proposed associations, and by consequence the project itself, would hold together only if the different entities concerned (electrons, catalysts, industrial firms, consumers) accepted the roles that were assigned to them. (Callon 1987: 93)

The success of these actor-worlds depends on the collaboration between the components that they assemble: these components might refuse to collaborate or their interests and positions might not have been appropriately assessed, or these scenarios might have been grounded in some misconceptions. For example, in the case of the electric car, the environmentalist social group did not take up the role of lobbying for a less polluting transportation system, while the construction of the battery proved to be more difficult than expected. When catalysts for the construction of the battery "refused" to collaborate and social groups proved to be weaker than expected, Renault engineers proposed another scenario in which the electric car was not included. The network of human and non-human actors described by the EDF engineers for the electric car did not concretize in real enrollments of actors.

Callon's concept of "actor-world" emphasizes that technology developers' "worlds" are simplifications of a complex reality that depend on their assumptions and interests. Furthermore, different actors might hold very different expectations, depending on their interests and assumptions. The success of technology developers' expectations depends on other actors' willingness and ability to collaborate. These considerations illuminate the case of the Nanopil. Several actors are mobilized in BIOS engineers' expectations of the Nanopil – it is not just nanowires, pumping systems and computer chips that figure in the Nanopil's developers-world. By mobilizing the context of national screening programs, the government and the Ministry of Health are also enlisted in the vision, a monitoring role is attributed to the doctors and to people over the age of 50, and a system of remote care has to function to allow the communication of the pill's test result to the doctor. The success of the screening program depends on the radio-signaling system to send the result to the user, and on the installment of a social infrastructure in which the doctor receives the message and communicates (it) to the patient. These expectations on the Nanopil are consistent because they build a context in which all the entities mobilized are linked up. But how to go about assessing the plausibility of these expectations?

De Laat's aforementioned "fictive script" analysis provides some methodological tools with which to address this question. It devises three exercises to make fictive scripts explicit and to describe the users, the actors, and the roles that are assumed in emerging technologies. Because they are based on expectations rather than on material objects, these descriptions have a fictional character and have to be considered as "thought experiments" (2000: 194).

The first exercise is a "100 % thought-experiment". It involves describing how the world would be if the emerging object would deliver exactly as expected. By employing this exercise, "the general characteristics of the object in question" can be quickly grasped and described together with the "barriers" that exist in today's world and that constrain the realization of this object; for example, if the object is the battery for an electric vehicle, this experiment would lead to the statement that gas stations should deliver electricity, which is not the case today. The second exercise is a "black boxing" experiment. It involves black-boxing the object as it is conceived by its designers and analyzing it as an "input–output device". In this way, the actors connected to this device can be identified; for example, Renault engineers building the battery in electric vehicles or the users of this vehicle. The third thought experiment involves analyzing the "relationships between actors incorporated into the object". In this way, a description can be offered of what is expected from future *users*. Taking again the electric vehicle as an example, de Laat explains that

> If the battery is designed as a single module, it may be rapidly exchanged for a new, fully charged one. If however it is divided over different places in a car (which for instance was the case with the first electric Peugeot's 106) it necessarily will have to be recharged at home or at an electricity station (ibid.: 194).

These three exercises are complementary and "provide a preliminary description of the future world the object defines" (ibid.: 195). Such descriptions enable a comparison between different anticipated futures as well as an eventual deliberation on emerging technologies.

These scripting experiments were originally elaborated with the purpose of enhancing actors' reflexivity on structural and social difficulties and to facilitate the process of innovation. The concept of "fictive script" has also been applied by (Boer et al. 2009) with the aim of bridging the gaps between promises of emerging technologies and "the realities of product development". Different "actor-worlds" are reconstructed in this methodology. Then, the different expected networks of actors and enrollments are mapped in order to point out the gaps in their expectations of the involved actors as well as in their roles and relationships. These conceptual and methodological tools offer some guidance when addressing the question of assessing the plausibility of expectations concerning an emerging technology.

4.3 Research Design

Nanopil (NP) developers hold scenarios of how this device will function in a broader social context. How are these assumptions inscribed in the technology's design? And to what extent do these scenarios match existing users and practices? How plausible are the visions held by the Nanopil developers concerning the use of this device in a social context when compared with the "actor-worlds" of other actors? These questions have been addressed in different steps.

First, we can reconstruct the actor-world of NP developers. As the overarching goal of my study is to assess the plausibility of the visions that the NP will realize some values (or a desirable world), I focused on the roles that are ascribed to users in defining the actor-worlds. This was done in several ways during the laboratory engagement presented in Chap. 3. Semi-structured interviews were partly based on the three thought experiments described above. I explicitly asked researchers to describe a world with the Nanopil as they envisioned it. In probing their answers,

4.3 Research Design

I asked them to define the actors involved in this world, along with their role. I also solicited them to reflect on the users and the tasks that they are expected to perform.[8] Because the inventor of the Nanopil is an oncologist, this type of interview was also conducted with him. This exercise allowed me to: (1) construct some fictive scripts of the Nanopil; (2) map a network of actors, as expected by the developers of the Nanopil.

The next step was to contact and interview these actors. In chronological order, I interviewed: one of the founders of the Dutch Institute for Prevention and Early Diagnostics (NIPED) who was also involved in the Nanopil patent application; an officer of the Dutch Health Council (Gezondheidsraad, GR) who supervised the Council's report on the necessity and availability of colorectal cancer population screening in the Netherlands; the officer of the Dutch Ministry of Health Welfare and Sport (VWS) who participated in the committee preparing this report; a staff member at the National Institute for Public Health and the Environment (RIVM), a specialized Dutch government agency in charge of the implementation plan for national-screening programs; a gastroenterologist[9] at the Erasmus Medical Centre, Rotterdam; the head of the oncology department at the Vrije University Medical Centre, Amsterdam; a General Practitioner; the contact for the patient group for intestinal cancer at the SPKS, Stichting voor Patiënten met Kanker aan het Spijsverteringskanaal (Foundation for patients with cancer of the digestive tract). The aim of these interviews was to (1) collect details about the current social practices that the Nanopil developers expect it to be part of and; (2) to ask those actors who might be affected by Nanopil to assess the plausibility of NP developers-world from their situated perspective. It should be noted that all these actors are not neutral judges; instead they should be seen as "stakeholders". They had some interest at stake in their judgment of the plausibility of expectations on the Nanopil.

Furthermore, the interviews with this variety of actors left room for digressions. I first invited them to describe their position and their role in the practice of colorectal cancer diagnostics, screening, management, prevention, information, etc. For example, the gastroenterologist was asked to describe the current practice of colorectal cancer diagnosis, while the RIVM researcher was invited to depict current screening implementation. In order to gain a better understanding of how the diagnostic practice for colorectal cancer is performed (Mol 1998), I also conducted participant observation in the colonoscopy room for one day at the Vrije University Medical Centre, Amsterdam. Furthermore, document analysis was conducted in order to understand the history of population screening programs in the Netherlands, as well as patients' perceptions of the current diagnostic and screening practice.

During the interviews, actors were invited to elaborate on their expectations regarding future trends in their field. Then, the Nanopil was added to their "world".

---

[8] This was not done via a strict protocol; instead the interviews were conducted in a conversational way. In some cases, these questions were not asked in one interview but on different occasions, including during my observation of their experiments, when judged appropriate.

[9] The gastroenterologist was a member of the GR committee working on the colorectal screening and an organizer of one of the feasibility trials for the screening tests for colorectal cancer.

While some of these actors were aware of the NP project, others were not. In those cases, I explained the working principles of the device and, building on my previous conversations with engineers, I discussed subsequent questions on more technical details. I also presented the "fictive scripts" constructed by and with the technology developers. Inspired by de Laat's experiments, respondents were asked to "quickly grasp the general characteristics of the object in question, and especially the 'barriers' existing in today's world to realize the object of concern" (de Laat 2000: 194). In doing so, special attention was given to the expected "antiprograms" of the users and to the relationships between actors that are inscribed into the NP (what does it pre-scribe, allow, forbid, stimulate, etc.). In addition to the barriers, the opportunities that actors associate with this emerging technology were investigated.

In the following section, it will be highlighted how this study contributes to addressing the questions about the plausibility of expectations surrounding the Nanopil. Since close discourse analysis was not a goal of my data collection, I will not always provide exact quotes although I do indicate the source of these considerations.

4.4 The Nanopill: Tales of an Emerging Practice

4.4.1 Nanopil Designers-World

Let us consider again the script of the video introduced in Chap. 3. The Nanopil is a micro-system that can perform elaborate analyses of biological samples. This system is small enough to be contained in a capsule. The capsule will travel through the intestine, collect some intestinal fluid, and detect abnormal DNA through changes in the electrical current of nanowires. Then, the pill will inform the user of the result: "The result of the measurement can then be sent via radio signals to a receiver (e.g. a mobile phone) that can send the result directly to the doctor" (Berg 2009). This script is included in a broader vision of portable devices for health diagnosis and monitoring:

> There are great expectations of micro-fluidics and Labs on a Chip. An important area is that of biomedical applications. You can think of a variety of Point of Care systems that can be used both for early diagnosis and patient monitoring. The wireless network-communication plays a major role. Thus, for example, an iPhoneMed unit could be developed that links the Lab on Chip Devices to the mobile phone. (ibid., *my translation*)

According to its inventors, the Nanopil functions as a screening test for colorectal cancer. This means that this device does not determine whether the user has colorectal cancer, but whether there are some conditions that are worthy of further investigation using techniques like colonoscopy. The NP is expected to replace other tests that are used for screening healthy people. In many descriptions and expectations, there is reference to the NP's possible use within a national screening program in which people over the age of 50 would be asked to take it (ibid.; Melchior 2009).

4.4 The Nanopill: Tales of an Emerging Practice

As I mentioned in the previous chapter, the Nanopil is a project initiated by an oncologist. According to the technology developers, Dr Pinedo provided the social context and need for the Nanopil.

> The problem in cancer treatment is that too many patients are too late. For example, you have complaints but you don't go to the doctor. Then, you have the doctor's delay, when the doctors misinterpret the complaints. Then you have the general practitioner's delay, like the waiting list. And so finally, when the patient is ready for treatment he cannot be cured. So the general thing, you see it with breast cancer, when I started 30 years ago, three out of four women died from breast cancer. Now only one out of four dies, because of early detection of breast cancer. The same is true for cervical cancer. Since you have the pap smear and a good checkup like in Holland, only a few patients die. But in many cancers you don't have these tests. For colon cancer you have the test of the stool but it is not very reliable. Often you have false positives and false negatives. So many people will still be lost because they are not tested. Also people do not like to bring the stool to the lab (Interview on April 12, 2010).

The brain-father of the Nanopil envisions the Nanopil as a tool with which people can monitor their health. He stresses that the Nanopil is not aimed at patients entering the doctor's office with a symptom, but to healthy people or "clients": "We are not talking about patients but about normal human beings", he stresses. He also points out that as a society, "we are going towards the age of prevention" and the Nanopil is the tool of the future for engaging in preventive practice. He envisions that Nanopil could be used by healthy individuals in the privacy and comfort of their home as a component of a prevention and self-monitoring routine.

Engineers involved in the development of the NP frequently refer to Doctor Pinedo's observations in order to justify their assumptions on the user and context of use. However, as the previous chapter has shown, conditions and contexts of use are also inscribed in the artifact during the research practice. This is the case for the laxative that enables the capsule to travel in the intestinal tract. Similarly, the sampling and analyzing operation is expected to be much longer than the 1 min shown in the video. Researchers expect that the whole operation will take between 7 and 15 h. Also, the original design of the NP was based on a Blue-Bolus result communication through the release of blue-dye coloring the stool. The program of action of this design choice had the individual or client as the only witness of the test result. However, the existing state of the art (e.g. the PillCam) and the technical challenges of including a blue-dye pumping system in the pill have led engineers to explore another system, namely a wireless radio-signal. In the program of action inscribed in this design, the results are communicated not only to the client but also to his/her doctor. Furthermore, the client will have to wear a receiving belt that can amplify the radio signal from the pill and send it to an external device such as a mobile phone or a computer.

Other aspects, not directly linked to the functioning of the artifact, shape the fictional description of the Nanopil's context of use. For example, a central element in the description of the NP is that it will be a cheap diagnostic device. However, as reported in the previous chapter, microchips are cheap only when they are produced on a large scale. Therefore, the description of the NP as a cheap diagnostic device assumes that it will be sold in large quantities.

Finally, a certain behavior is assumed by the users: the users will clear their intestine by drinking liquid laxative the day before the NP ingestion (as when patients prepare for a colonoscopy). They will swallow the pill and wait for it to perform the in vivo test. In the case of the radio-signaling configuration, after a few hours they will attach a belt and will keep their (smart) mobile phone nearby. In these descriptions, different contexts, actors, institutions and conditions are imagined to cooperate. The following sections, based on interviews with other actors and document analysis, will thicken, articulate and/or question the technology developers' descriptions of the context of use, the practice and the user of the Nanopil.

4.4.2 Comparing Actors' Worlds: Current Screening Practice and Future Trends

The Dutch Health Council report on colorectal cancer screening is mobilized in the Nanopil developers-world. The Dutch Health Council officer who supervised the report spoke about the story of the report. To date, the Netherlands has successfully implemented two national screening programs: breast cancer and cervical cancer. These screening programs have proven successful in decreasing the mortality rate for these two diseases. The debate on the national program for screening in the Netherlands was triggered by the results of a European survey in which citizens' awareness about colorectal cancer was explored. In comparison with other Europeans, Dutch citizens appeared to be less aware of the danger and symptoms of colorectal cancer. The results of this survey initiated a debate on the introduction of a screening program in the Netherlands.[10] Indeed, in 2007 the Dutch Minister of Health, Welfare and Sport (VWS) asked the Health Council to produce a report on the necessity and availability of colorectal cancer screening in the Netherlands. The expert committee set up by the Health Council judged unanimously that a screening program for colorectal cancer was desirable. Subsequently, several pilot studies began to test the sensitivity and specificity of available tests for CRC.[11] The iFOBT (immunochemical fecal occult blood test) was selected by the committee as the most suitable test for a population screening program. This self-test (conducted by a healthy individual by collecting stool samples and sending them to the laboratory) would reveal the presence of occult blood in the feces, which is a symptom of a potentially bleeding tumor. Although the test has a relatively low sensitivity (65 % against the 97 % for the colonoscopy test), an analysis of its costs and benefits

---

[10] While in other European countries like France, the UK and Germany there was some form of national program for population screening for CRC, this was not the case in the Netherlands.

[11] Namely, gFOBT (the "guaian" fecal blood test), iFOBT (an immunochemical fecal blood test, specific for human blood and more sensitive than the guaian), colonoscopy (considered as the golden standard and always performed in any trial as control), CT colonography (or virtual conoscopy), sigmoidoiscopy (similar to a colonoscopy, but examining only the final part of the colon), molecular test (detecting specific biomarkers in stool or blood).

during the trials showed it to be the best available solution at the moment (Health Council of the Netherlands 2009).[12] A subsequent colonoscopy would however be necessary to diagnose the eventual presence and type of tumor.

While the Dutch Minister of Health at the time endorsed the final opinion of the Health Council, he highlighted the difficulties of implementing such a screening program in the short-term in a letter to the Parliament in November 2009; reasons included the lack of immediate capacity, in terms of staff and infrastructure, to guarantee a diagnostic examination (i.e. a colonoscopy) to every individual with positive results.[13] While working on increasing the capacity of training nurses and other clinical personnel to conduct endoscopic investigations, the RIVM (the National Institute for Public Health and the Environment)[14] defined the implementation plan.

How would the Nanopil fit in this context? The officer of the Dutch Ministry of Health, Welfare and Sport responsible for the colorectal cancer screening program envisions it as a possible replacement for the iFOBT.

> With the iFOBT, we invest in an infrastructure that we can easily adapt when we get a better test. Because we know that we will have a better test in the foreseeable future.

She goes on to explain that the implementation plan of this program was based on the idea of creating an infrastructure that is flexible enough to adapt to newer, more reliable self-tests than the iFOBT. However, the description of the screening practice provided by the staff member at the RIVM sets some conditions on this vision.

With respect to its implementation, colorectal cancer screening is different from the other existing screening programs because it involves a larger number of people (men and women over 50) and it implies a self-test. Indeed, while the smear test for cervical cancer and the mammography for breast cancer are conducted by medical personnel in a medical setting, the iFOBT requires that the stool sample collection is carried out by the healthy individual undergoing a screening program (or "screenee") at home. The sample is then analyzed in the laboratory. In the currently running national screening programs, the general practitioner (GP) is a focal point. In case of an unfavorable test the GP receives the result of the test from the screening organization, informs the screenee and refers him or her to hospital. For that reason the GP is informed several days before the screenee receives the letter with the result. In the case of cervical cancer, the GP is also responsible for performing

---

[12] In order to evaluate the introduction of a population screening program, the Health Council drafted a normative framework in 2008 based on the ten principles of screening that were formulated by Wilson and Jungner in 1968. This normative framework comprises five criteria: (1) screening is directed at important health problems; (2) screening results in health gains or other benefits of the test subjects in question; (3) the screening method is reliable and valid; (4) participation in screening and follow-up examinations is based on an informed and voluntary choice; (5) efficient use is made of resources (Health Council of the Netherlands 2009: 27).

[13] On May 25, 2011, the Dutch Minister of Health, Welfare and Sport (VWS) decided to introduce screening for bowel cancer. The programme is expected to start between 2013 and 2019 (http://www.rivm.nl/Onderwerpen/Onderwerpen/B/Bevolkingsonderzoek_darmkanker)

[14] See http://www.rivm.nl/en/

the smear test. As the RIVM staff member explained to me (at the time of the interview), the ongoing debate at RIVM centered on whether (in the future CRC screening program) the GP should also play a role in referring a screenee – in the event of a positive result – to the hospital. For reasons of quality and in order to speed up the appointment process, this task could probably be performed better by the regional screening organizations. In fact, in describing the implementation plan for screening programs, he pointed out how the Centre for population screening within the RIVM organizes the "logistics" and the "coordination" of the program. These screening programs aim at balancing "health improvement" and "autonomy" (as self-determination) for the involved individuals. Within this perspective, a screening program in which individuals are the only recipients of the result of the test is unconceivable because it would be impossible to "organize the screening". This sets some conditions on the use of the Nanopil for population screening.

As explained above, the Nanopil developers are exploring two alternative designs for informing the user of the test result. These two designs are not equivalent in view of the use of the Nanopil as a test in a population-wide screening program. If the Nanopil communicates the result via radio signal and if the notification can be sent not only to the user's device (i.e. a mobile phone), but also to the GP's, it could be received instantaneously by the doctor and the user. In the blue bolus pill, instead, the subjects are informed of the result of the pill via the release of the blue dye visible in their feces. In the "blue-bolus pill" scenario, the screening subjects are the only witnesses and monitoring by medical personnel is not possible. The program of action inscribed in the "blue-bolus pill" seems not to be compatible with the program of action inscribed in the screening programs. The details introduced by the RIVM representative enrich the image of the current population screening practice and challenge the consistency of the Nanopil developers' world.

The scenario is different in the case in which the Nanopil is conceived in a "self-test" context; the blue-bolus pill is considered as absolutely plausible for such a context. One of the founders of NIPED (the Dutch Institute for Prevention and Early Diagnostics) and an applicant for the Nanopil patent explores the possibility of having the "blue-bolus pill" as a component of the Prevention Compass. This is a prevention kit for personalized risk profiling, developed at NIPED.

> The Prevention Compass [is] a patented, web-based, knowledge and decision support (KDS) system for evidence-based personalized prevention [...] It aims to empower professionals and individuals by delivering state of the art scientific insights while facilitating task-delegation, health education, self-management, quality assurance and shared decision making.[15]

This "@home lab-box" provides the tools for the home collection of heterogeneous data such as questionnaires, biometrical data, and biological samples that will subsequently be sent to the laboratory for analysis. The web-platform integrates these data and determines a personalized risk profile for several health conditions (cardiovascular risk, mental disorders and colorectal cancer). The main idea is that

---

[15] See: http://www.preventiekompas.nl/HomepageRedesign.aspx?menu_id=9&page_id=136

this prevention platform should be personalized, that is, able to collect sufficient information from the individual and to provide a personalized response concerning the risk profile and the type of intervention required. Such a kit is expected to be sold directly to the consumer or to be used in healthcare practice as a management tool.[16] In this context, the blue-bolus design is a viable option.

The Nanopil does not constitute a more reliable test that can replace the iFOBT in the future and simplify the screening procedure. On the contrary, the final design of the Nanopil will contain a script that will allow some social configurations and exclude others. Currently, several design options that hold different fictive scripts co-exist. Some of them are more plausible than others not only with respect to their technical feasibility ("the radio-signaling system is easier to reproduce because it already exists on the market/it is easier to miniaturize"), but also because of the social context in which they are expected to operate. In light of the program of action held by the screening organization, the expectations that Nanopil will be used in a screening program are plausible only if the developed design for the pill involves a communication system that sends the result to a monitoring structure (i.e. the GP or the regional screening Center).

4.4.3 Users' Preference and Resistance

The designers have a certain idea in mind with respect to how the users behave, what they want, how they look. These assumptions contribute to shape the technology. However, these assumptions can be misleading, failing to address the real users, who differ with regard to their fears, preferences, education, interests and values. In order to assess the plausibility of the expectations of users' acceptance of the artifact, the fictive script inscribed in the design can be compared with their preferences and interests. In this way, deviance of users' behavior from the expected one and resistance to the program of action inscribed in the pill can be taken into account.

Although agreeing in principle with the importance of prevention, the general practitioner and the gastroenterologist responded to the "home-kit" vision by describing some aspects of the current practice of colorectal cancer diagnostics that cannot be reduced to a self-test. In the view of the gastroenterologist, the practice of making a diagnosis is much more than just performing a test:

> Making a differential diagnosis is not different from making a hypothesis. When you make a hypothesis, you take a few things into account. One is the *probability* of your diagnosis: the more probable, the higher you put it on the list of the differential diagnoses. The second is the clinical *relevance* of the diagnosis: if a physical condition is totally irrelevant, you put it low on your differential diagnosis. And the third is the *ease* and *certainty* with which you can test it. So sometimes you can have a wonderful option in your hypothesis, but if you don't have any test to prove or disprove it, or if it is a very difficult test, it is not very useful.

---

[16] In fact the Prevention Compass motto is: "knowing, measuring, doing" (weten, meten, doen).

So if you take this into account, and you go back to the patient with blood in the stool, a differential diagnosis of colorectal cancer, if the patient is 60 years and has blood in the stool is quite *probable*. It is also *relevant* to the patient to prove or disprove. But the blood in the stool test is not a good test for the symptomatic patient: the test may be negative, but there is still a chance that the diagnosis is correct. So then the test doesn't help in that setting. A doctor seeing a patient with symptoms should not test the stool. What he can do is test for anemia, and do some type of imaging of the colon with which you can truly corroborate or exclude the diagnosis of cancer (*My emphasis,* interview on June 9th 2010).

When a patient walks into the doctor's office with some symptoms, a screening test is not an option. "The history of a patient says much more than any test", echoes the GP. The verbal interaction between the doctor and the patient plays a central role at this stage: the patient's family history, the description of the symptoms, and observation help the doctor in choosing one diagnostic tool over another. Moreover, it helps the GP to guide the patient in making decisions. For example, in the colonoscopy room where I carried out participant observation, the gastroenterologist carefully read the history reported by the GP, and based on that history, made decisions related to how to perform the colonoscopy (whether to use anesthetics, the desired width of the scope, etc.). The NIPED-world seems to conflict with the gastroenterologist and general practitioner's world.

Furthermore, the Nanopil developers-world conceives this device as a more userfriendly alternative to the existing screening test; this is because the Nanopil does not require people to collect a sample of their own stool in order to send it to the lab. In this way, technology developers assume that the lack of early screening is due to people's unwillingness to collect their stool sample. However, the GP and the patient advocate do not view the stool collection as a problem; rather they point to the lack of awareness on the part of the Dutch population regarding the risks and symptoms of colorectal cancer as a reason for late intervention.

Furthermore, the literature shows that the bowel preparation preceding the examination is often considered to be the most burdensome and unpleasant aspect of the colonoscopy (Ristvedt et al. 2003; van Gelder et al. 2004; Beebe et al. 2007). The laxative has the purpose of clearing the intestine of feces in order to allow good visibility of the intestine walls during the endoscopy. Patients complain that the consumption of the laxative compels them to spend the day before the colonoscopy ingesting 5 l of liquid laxative and "sitting on the WC". A special diet, poor in fibers, has to be ingested and the treatment causes nausea and sickness. A similar "bowel preparation" condition is embedded in the NP design, in order for the pill to freely move through the intestine and sample intestinal fluid. The user is expected to ingest a laxative and prepare the bowel before swallowing the pill. In this respect, the pill does not offer an easier and more user-friendly alternative to the FOBT since it requires the user to ingest the laxative.

Finally, the patient advocate and the general practitioner emphasize new aspects. Mistakes, fears, and "irrational" behaviors play a central role in the current diagnostic practice and are even more noticeable in the case of self-tests. The GP, for example, remarks that self-tests need to be extremely clear and easy to interpret. This remark is based on his experience with people's interpretation of the pregnancy self-test: although the result notification seems to be rather straightforward (+/−), users are often unable to interpret it correctly.

4.5 Conclusions

Expectations on emerging technologies are projections of a future world in which the new artifact is described within a social context. In the visions of the Nanopil we saw how technology developers tend to sketch coherent worlds in which the artifact, user groups and other stakeholders entertain simple and clear relationships. Assessing the plausibility of expectations about use is both a critical and a constructive activity. By discussing the envisioned technology with other actors, it is possible to identify where developers' visions collide with other actors' views and beliefs. But it also adds new perspectives and thickens the initial visions by exploring how current practices might change and adapt, in addition to new problems that might surface.

The concepts of script and actor-worlds suggest that technology developers hold some assumptions about the prospective users and other actors around a certain innovation. As the fictive script analysis suggests, these constructions can be analyzed by shifting from one actor-world to another in order to map the displacement between them. This can be done through involving different potential social actors and users in future-oriented experiments and sketching the context of use, as well as the roles of these different actors; these conceptual tools have guided my research design.

Following the construction of Nanopil developers-worlds, I emphasized that some entities in such worlds need more elaboration through the exploration of other actor-world. Additional details on a social practice, like the population-screening program, exclude some visions of the Nanopil and make others more plausible. When compared with the world of the National Institute for Public Health and the Environment (RIVM), the expectation of the use of the Nanopil within the context of a population-screening program is plausible only if the Nanopil features a system that allows the screening Centre to monitor the result. The scenario of the blue-bolus pill is not consistent with that of the population screening program. The assessment of expectations on the Nanopil has also highlighted the potential for resistance and controversies, as for example in the case of general practitioners or screenees required to ingest liquid laxative.

The analysis of the expectations of the Nanopil followed the double strategy of thickening and situating. Details were added to the picture, by exploring the situated perspective of actors in the practice in which the emerging technology is expected to operate. Furthermore, the assessment of the plausibility of expectations of the Nanopil was situated in the perspective(s) and practice(s) of the actors I interviewed. As pointed out, these actors are not neutral judges, but "stakeholders": their interests played a role in their judgment of the plausibility of expectations on the Nanopil. Therefore, their assessment and description should not be taken at face value. I have not engaged in the assessment of the plausibility of these views; rather my interest was in positioning one world in relation to the other.

One final remark: it is only upon leaving Paris for Africa – where she is confronted with the actual use of the artifact – that the significance of some technical components occurs to Akrich. "It was only in the confrontation between the real user and the projected user that the importance of such items (…) came to light" (1992: 210). I share this observation: it was only following the encounter with the current cancer diagnostic practice and the expectations of potential users that some "technical" components of the Nanopil stood out. For example, only during the interview with the RIVM researcher did I understand the extent to which the choice between a radio-signaling and a blue-bolus system affects the potential context of use. The comparison between the expectations of the Nanopil's developers within the laboratory and the expectations of other actors highlighted the significance of some "technical components" and design choices. In this sense, the story narrated in the previous chapter was written a posteriori, following my interviews with other actors and potential users. Similarly, the story narrated in this chapter aims to address the question of the plausibility of the visions on the desirability of the Nanopil. The next chapter will explain how this can be done.

References

Akrich, M. 1992. The description of technological objects. In *Shaping technology building society: Studies in sociotechnical change*, ed. W. Bijker and J. Law. Cambridge, MA: MIT Press.
Beebe, T.J., C.D. Johnson, S.M. Stoner, K.J. Anderson, and P.J. Limburg. 2007. Assessing attitudes toward laxative preparation in colorectal cancer screening and effects on future testing: Potential receptivity to computed tomographic colonography. *Mayo Clinic Proceedings* 82: 666–671.
Callon, M. 1981. Pour Une Sociologie Des Controverses Technologiques. *Fundamenta Scientiae* 2(3/4): 381–399. Mines ParisTech París.
Callon, M. 1986. The sociology of an actor-network: The case of the electric vehicle. In *Mapping the dynamics of science and technology: Sociology of science in the real world*, ed. M. Callon, J. Law, and A. Rip, 19–34. London: MacMillan.
Callon, M. 1987. Society in the making: The study of technology as a tool for sociological analysis. In *The social construction of technical systems: New directions in the sociology and history of technology*, ed. W.E. Bijker, T.P. Hughes, and T.J. Pinch, 83–103. Cambridge, MA/London: MIT Press.
de Boer, D., A. Rip, and S. Speller. 2009. Scripting possible futures of nanotechnologies: A methodology that enhances reflexivity. *Technology in Society* 31(3): 295–304.
de Laat, B. 1996. *Scripts for the future: Technology foresight, strategic evaluation and sociotechnical networks: The confrontation of script-based scenarios*. Thesis (Ph.D.), Universiteit van Amsterdam.
de Laat, B. 2000. Scripts for the future: Using innovation studies to design foresight tools. In *Contested futures: A sociology of prospective techno-science*, ed. N. Brown, B. Rappert, and A. Webster. Aldershot: Ashgate.
Health Council of the Netherlands. 2009. *A national colorectal cancer screening programme*. The Hague: Health Council of the Netherlands. Publication no. 2009/13E.

References

Hughes, T.P. 1983. *Networks of power: Electrification in Western society, 1880–1930*. Baltimore: Johns Hopkins University Press.

Hughes, T.P. 1987. The evolution of large technological systems. In *The social construction of technical systems: New directions in the sociology and history of technology*, ed. W.E. Bijker, T.P. Hughes, and T.J. Pinch. Cambridge, MA/London: MIT Press.

Latour, B. 1992. Where are the missing masses? Sociology of a few mundane artefacts. In *Shaping technology, building society: Studies in sociotechnical change*, ed. W. Bijker and J. Law, 225–258. Cambridge, MA: MIT Press.

Melchior, M. 2009. Doctoren met nanotechnologie. *Medisch Contact* 64(49): 2032–2035.

Mol, A. 1998. Missing links, making links: The performance of some artheroscleroses. In *Differences in medicine: Unravelling practices, techniques and bodies*, ed. A. Mol and M. Berg, 141–163. Durham: Duke University Press.

Oudshoorn, N., and T.J. Pinch. 2003. *How users matter: The co-construction of users and technologies*. Cambridge, MA: MIT Press.

Pinch, T.J., and W.E. Bijker. 1987. The social construction of facts and artifacts: Or how the sociology of science and the sociology of technology might benefit each other. In *The social construction of technical systems: New directions in the sociology and history of technology*, ed. W.E. Bijker, T.P. Hughes, and T.J. Pinch. Cambridge, MA/London: MIT Press.

Ristvedt, S.L., E.G. McFarland, L.B. Weinstock, and E.P. Thyssen. 2003. Patient preferences for CT colonography, conventional colonoscopy, and bowel preparation. *American Journal of Gastroenterology* 98: 578–585.

Silverstone, R., and E. Hirsch. 1992. *Consuming technologies: Media and information in domestic spaces*. London/New York: Routledge.

van den Berg, A. 2009. De kunst van het kleine, in Brinskma and van den Berg, De kunst van de wetenschap, Redevoeringen 48ste Dies Natalis, Universiteit Twente.

van Gelder, R.E., E. Birnie, J. Florie, et al. 2004. CT colonography and colonoscopy: Assessment of patient preference in a 5-week follow-up study. *Radiology* 233: 328–337.

von Hippel, E. 1988. *The sources of innovation*. New York: Oxford University Press.

Chapter 5
The Good in the Pill. Assessing the Plausibility of Visions of Desirable Worlds

> *There is no one best way to paint the Virgin; nor is there one best way to build a dynamo. Inexperienced engineers and laymen err in assuming that there is an ideal dynamo toward which the design community Whiggishly gropes. Technology should be appropriate for time and place; this does not necessary mean that it be small and beautiful. (Hughes 1987: 68)*

Abstract Building on the analyses provided in Chaps. 3 and 4, this chapter establishes that general claims regarding the desirability of an emerging technology appear to often draw on a superficial unifying rhetoric of supposedly shared values. The discourses on the desirability of the Nanopil implicitly or explicitly refer to a number of values – autonomy, care, comfort, efficiency – that the technology is claimed to promote. The analysis demonstrates the ambivalence and contradictions inherent in the expectation that a technology "offers a solution for a social problem", "addresses a need" and "improves our current practice". It does so by building on frameworks developed in the field of philosophy of technology and within the "Vision Assessment" approach. Such frameworks bring forward the moral connotation of possible design choices and the latent ethical controversies in stakeholders' normative positions, as well as potential technology-mediated changes in the current moral landscape and value framework.

Keywords Vision assessment • Technical codes • Techno-moral change • Technology mediation

5.1 Visions of Desirable Worlds

Expectations of emerging technologies describe innovative artifacts and project the way in which they will be used in a social practice as something "good" or "desirable":

> "Doctors will be able to use the pill in five to ten years only "says Pinedo." But then you have something very special: early diagnosis without the use of stool samples. With the signals revealing the results, physicians may determine the need for a colonoscopy. Patients

no longer need to stand with a jar of stool in the corridor of the hospital laboratory for an initial test. If you can determine whether people have a high risk of colon cancer with a pill, then affordable and less burdensome colon cancer screening is possible." (Melchior 2009, *my translation*)

Discourses around the Nanopil often highlight how and why the Nanopil is desirable (or good) for individuals and society. These expectations describe the "good" in the pill: how the Nanopil provides a desirable means to address a social need, that is, detecting colorectal cancer at an early stage.

Dr Pinedo, oncologist and "father" of the Nanopil, explains[1] that the need for a Nanopil occurred to him during his observation of existing screening practices. In particular, he refers to the decrease in mortality rates for cervical and breast cancer following the invention of reliable screening tests and their introduction into a large target population. This observation reinforced Pinedo's idea that more screening tests of this sort are needed to intervene earlier and to increase the probability of success of eradicating a growing tumor at an early stage. In order to be effective, these screening tests have to be reliable in determining the risk of cancer and affordable in order to be broadly distributed in society. Moreover, in order to be successful, screening practices must be comfortable and not taxing for the patient.[2]

Dr Pinedo emphasizes that, in the case of screening, the people who use the test should not be viewed as patients, since they are monitoring their health without experiencing any symptoms. The autonomy of these asymptomatic clients is a priority in the oncologist's opinion. The test should also be easily incorporated into clients' routine and should respect their cultural background. This seems to imply that one has to adapt the screening practice to the local culture. For example, Dr Pinedo describes the good practice employed in the introduction of mammographic screening in South American villages. In order to come to terms with the local culture of shame around such an invasive test requiring women to have their breasts monitored, the screening practice was organized via a specially-arranged transportation system. By driving the "screenees" to other villages, where their anonymity was preserved, the screenees were spared the "shame" of being exposed to the prying eyes of the local community. As in the case of breast cancer screening, also in the case of colorectal cancer screening there seem to be a kind of taboo to be circumvented.

As Prof van den Berg explains, the available methods for detecting colorectal cancer are painful and expensive, like in the case of endoscopic investigations. When simple screening tests are available, like the Fecal Occult Blood Test (FOBT), they are not user friendly. The latter is uncomfortable because it requires the screenee

---

[1] Interview conducted on April 12, 2010.
[2] See also a previously quoted expectation by Prof van den Berg, the head of the BIOS group developing the Nanopil: "Colonic cancer is one of the most common cancers in people over the age of 50. The Dutch Health Board has already advised endoscopic or colonoscopic screening for people in this age group. But this is a painful and uncomfortable experience. What is more, it presents a logistical nightmare. And nothing is found in 95 % of cases. What we need is a simple first-line test. The only alternative at present is a faeces test, but eventually, a nano-pill will provide a much more patient-friendly alternative" (Berg 2009).

to collect a stool sample and send it to the laboratory to check for blood traces. This is "not the best hobby" as the oncologist remarks, or, as one of the BIOS researchers exclaims to highlight the old-fashioned character of such sample collection, it is "a medieval practice!".[3] The Nanopil proposes a "technological" solution to this discomfort. The promise is therefore to provide the users (both the screenee and the GP) with a means to effectively monitor bowel condition in an easy and comfortable manner. The NP offers a way for people to monitor their health and to detect abnormal statuses at a very early stage since it tests for the molecular causes of cancer and it is a "clean", non-"medieval", modern way of testing and receiving results.

The promises of Nanopil emphasize cost-effectiveness, the increased chance of saving human lives, the increased autonomy of the user, and a decrease in discomfort as valuable expected outcomes of the introduction of the Nanopil in the colorectal cancer screening and diagnostic practice. In Chap. 2, I pointed out the strategic and rhetorical character of expectations. Mobilizing values, placing emphasis on needs and proposing desirable futures prompt public support and interest around a new technology. However, I have also shown that technoscientific expectations guide societal decisions in the direction of envisioned desirable worlds. In the visions of the Nanopil, for example, a linear link is made between the technology, the clinical practice and the attainment of these desirable outcomes. The Nanopil is expected to be a means for promoting and protecting some values, the desirability of which is taken for granted by the Nanopil developers. In Chap. 2, I gave a *prima facie* argument that *in order to assess the desirability* of emerging technologies it is not enough to articulate the normative content of these visions. Rather, *the plausibility of these visions has to be assessed*. As I argued, three questions need to be addressed:

1. How likely will the expected *artifact* promote the expected values?
2. To what extent are the promised values *desirable for society*?
3. How likely is that a technology will *instrumentally* bring about a desirable consequence?

To address these questions, I have critically analyzed the expectations of the technological artifact as well as the expectations of its context of use, respectively in Chaps. 3 and 4. On the basis of these analyses, I address here the questions about the plausibility of the Nanopil visions: whether the artifact is likely to realize the claimed values (Sect. 5.2), whether values are indeed considered valuable by everybody (Sect. 5.3) and, which other values might be affected (Sect. 5.3).

5.2 Different Expected Artifacts and Different Values

In his book *Alternative modernity*, Andrew Feenberg refers to technologies as "meaningful objects" (1995: 155). He explains that technologies have two types of meaning: (1) a function that the technology is supposed to perform (2) and some

---

[3] From an interview on April 15, 2010.

connotations that "associate technical objects with other aspects of social life independent of function" (Ibidem). For example, the automobile has a function of transportation and some connotation about the social status of the owner. Feenberg argues that, in new technologies that are still emerging, the function and connotations are not clearly distinguished and a potential technological object might have a variety of functions and connotations. Referring to a classic example in the Social Constructivist approach to technologies (Pinch and Bijker 1984), he explains how, in the early development of the bicycle, different designs coexisted depending on the connotation that users attributed to bicycling. In some cases bicycling was considered a competitive sport and therefore the "velocity" function was privileged. In other cases, bicycling was viewed as a means of transportation and therefore the "safety" function was the priority. These two bicycle designs are functionally and connotatively different. According to Feenberg, the "ambiguities in the definition of a new technology must be resolved through technical development itself" (1995: 156). In the process of technical development, one design choice will prevail and the struggle of meaning will be covered up (the process described by Latour 1987 as technological "closure"[4]). Such a process of "closure" is consolidated in a "technical code". For Feenberg, "technical codes define the object in strictly technical terms in accordance with the most general social meanings it has acquired" (ibidem). However, these codes are the result of some social struggles or negotiations, and the meanings, in terms of the functions and connotations that they convey, reflect the ideology of one social party.[5] For example, in the case of the bicycle, the accepted morality at the time, according to which women's dresses were long and fully covering, prevented them from safely biking the faster bicycle thanks to a wider front wheel. In this sense, current morality has a role in the consolidation of a technical code.

Also in the case of the Nanopil two alternative technical codes can be pointed out. Although "the" Nanopil appears to be a well-defined artifact in the public promises, the emerging state of this technology blurs this definition. As I explained in Chap. 3, engineers' expectations on *one* emerging technology might in fact entail a *number* of different competing designs. For example, developers propose two technical alternatives for how the user can be informed of the result of the test. According to the first one, in the event that an abnormal DNA state is detected in the intestinal fluid, the pill releases a blue dye into the bowel, which is visible in the users' stool, once they evacuate. According to the second alternative, the pill sends a numeric result via radio signal to an external device, namely a mobile phone or a computer. As explained in Chap. 4, the expected use of the pill within a population-screening program for colorectal cancer excludes the blue-dye system, because the result has to be supervised by a care provider in a screening context. If the developers opt for the blue dye solution, it is unlikely that NP will be considered to be an appropriate test for a population screening program by the Ministry or Health

---

[4] See Chap. 3 for an explanation of the concept of "technological closure".

[5] This analysis brings Feenberg to draw attention on the fact that the ideologies and power struggles are settled in the consolidation of a certain technological design.

5.2 Different Expected Artifacts and Different Values

Council in the Netherlands. In addition, alternative technical options might not only influence the expected social context of use of a future technology, but also its moral meaning. Let us see.

The alternative technical codes that surround expectations on the Nanopil assume different functions of the device. The apparatus can function as a self-test, the result of which is manifested to the user who observes the (blue) colored stool. Otherwise, it can function as a screening device controlled by a medical professional who receives the numeric result of the test on an electronic device, such as a computer or tablet. These different technical codes and functions also carry different "moral"[6] connotations. To clarify this, let's look at how the connotation of the "moral status" of the expected users of the Nanopil differs in each of the two functions attached to the device. In the case that the Nanopil has a self-test function, the users will voluntarily and periodically test themselves in the comfort of their own homes. In expecting so, users are understood by technology developers as human beings who value prevention and are able to take control of their health. The expected users consider health monitoring to be important and are willing to monitor their health. They are also physically, intellectually and economically able to monitor their health. The Nanopil is expected to empower these users' autonomy because it allows them to decide if, when, where and (partially) how to monitor their health condition.

By contrast, if the Nanopil is expected to function as a screening device, users will be part of a control-system. In this system, the public healthcare organizations will take care of them by offering them the opportunity of undergoing a colorectal cancer test, depending on their demographic or their risk-profile. The public healthcare system motivates users to test themselves by providing them with a comfortable, easy, and "clean"[7] test. In this case, it is not the autonomy of the test's user from the external care provider that is emphasized. Rather, the efficiency and the quality of the care process are privileged. The quality of the care process is strengthened through augmented control over the result by the medical practitioner via a telecare system, that is, a system that can provide care in the physical absence of care personnel (for example via a computer).

Claims about the desirability of the Nanopil mobilize values such as "autonomy", "efficiency" and "comfort". However, these values are promoted differently in alternative technical codes. As a self-monitoring test, the Nanopil is primarily expected to promote the user's autonomy in dealing with her health decisions. As a screening device, the Nanopil is primarily expected to improve the quality of care in colorectal cancer clinical practice by providing an efficient system for the early diagnosis of colorectal cancer. These two technical codes underpinning expectations on the Nanopil embed different conceptions of what is "good" care.

I use a broad definition of "value" as a "judgment of an individual or a certain community about what is important" in life or more specifically in a specific social practice. The comfort of a medical examination is a value in the sense that it is considered important, good or worth striving for within a certain community, society, or

---

[6] For a discussion of how the concept of "morality" is intended here see Chap. 1.
[7] NP is supposed to be a "clean" test as opposed to the Fecal Occult Blood Test, which requires the user to collect a stool sample and send it to the laboratory for analysis.

practice. It can be claimed that patients' privacy, screenees' autonomy, the robustness of diagnoses, and patients' care are all interrelated values in the current practice of diagnosis of colorectal cancer. However, if one looks at how these values are inscribed in different technical codes, these values may conflict. For example, if the Nanopil is used for screening purposes, the result of the home-test has to be monitored. This can be done via a radio-signaling system. During a public debate on the Nanopil[8] and a follow-up interview,[9] an advocate of the Dutch organization for patients with cancer in the intestinal tract[10] voiced her fear that such a system endangers the patient's privacy. Therefore, the home-screening test is expected to enhance the *autonomy* of users who can self-administer the test, but it does not necessarily protect users' *privacy*. Giving priority to the value of autonomy might also endanger the *effectiveness* of the screening and the *quality of care*. The GP I interviewed made this point: the remoteness of the test's practice does not allow the GP to face the patient in person during the delivery of the result, possibly leading to undesirable misunderstandings. If radio-signaling systems are avoided in order to secure the user's *privacy*, the users remain the only actors in the healthcare chain to know the result of the test. Their privacy is promoted, in addition to their autonomy, because they can decide to take no action in the event of a positive result (abnormality detected). However, the *efficiency* of the screening system is diminished because a control-system is left out with potential consequences for the quality of care. Depending on what functions and technical codes will be stabilized in the artifact, different values for the colorectal cancer diagnostic practice will be privileged.

The idea of morality as a "force field" (Swierstra 2010) can help explain this value-tension in different technical codes. In the force field of morality, "conflicting norms and values compete for hegemony". Similarly, the values of efficiency and comfort of a telecare system compete with the value of privacy of the patient. In order to use the Nanopil as an efficient screening system for colorectal cancer, the patient's private data has to be sent via radio-signal to an external receiver and from there onwards to a communication network of which the GP is part. Privacy, together with the value of interpersonal relations in care, are downplayed while the value of efficiency gains more force.

In this sense, paraphrasing Feenberg, we can say that expectations on emerging technologies are ambiguous with respect to their moral connotation. Eventually, some technological "closure" (see Sect. 3.3) occurs and one moral connotation will dominate the others. It is unlikely that the Nanopill will realize all the promised values equally. Some technical codes will prevail and with them some values will be privileged at the expense of others. In order to reflect on the desirability of emerging technologies[11] and engage in an ethical technology assessment it is important to make explicit and articulate the moral content of expectations at a stage in which the

---

[8] Debat Nanotopia: http://www.lux-nijmegen.nl/debat/nieuws/2010/06/10/goed-idee-niet-de-nanopil-gaat-er-komen-dat-zeker

[9] This interview was conducted on July, 1 2010

[10] Stichting voor patiënten met kanker aan het spijsverteringskanaal (http://www.spks.nfk.nl/).

[11] This is what I referred to as Desirability-Question (or D-Question) in Chap. 2.

technology is still fluid and a technical code has yet to prevail. I have done this by singling out the alternative expected designs of "one" technology. Instead of reflecting on the Nanopil and its desirability, different artifacts (NP1, NP2, NP3,....) are distinguished in these expectations. The moral content of these different expectations is then articulated with reference to the technology's expected context of use and function. The assessment of the plausibility of expectations of emerging technologies that I propose serves the purposes of an ethical technology assessment by diversifying expectations on the artifact, rather than clustering them together under an empty "black-box".

5.3 Plurality of Values Among Actors

Values, interests and ideas of what constitutes a desirable world vary among social actors. The case of the high-wheel bicycle once more provides a case in point.

> The using practice of the social group of "young men of means and nerve," that is, racing, showing off and impressing the ladies, constituted the macho machine, whereas the using practice of the social groups of women and elderly men, that is, touring, falling off, and "breaking limbs and bones", constituted the unsafe machine. The macho machine led to a design tradition with larger wheel radius, and the unsafe machine gave rise to a variety of designs with, for examples, smaller wheels, backward saddle, or smaller wheel in front. Thus different using practices may bear on the design of artifacts, even though they are elements of technological frames of nonengineers. (Bijker 1987: 172)

Members of a social group share one meaning of the artifact (or, "technological frame"). This meaning varies from group to group according to their interest, their role and value system. As pointed out within the Vision Assessment approach,[12] expectations on the future depict "the world described by someone who is asked why particular technologies are desirable" (Reuzel and der Wilt 2000: 53). In this sense, visions "reflect the values, worldviews and deep preferences of those who hold them" (Grin and Grunwald 2000: 11). According to this approach, diverging normative perspectives of different stakeholders should be explicitly articulated, standards and criteria of merit of a technology analyzed, needs spelled out, problem **definitions** discussed, and "desirable final states" envisaged. This articulation is important in order to assess the plausibility that a technology will have "desirable" consequences. But for whom are these consequences desirable? Diverging worldviews and normative positions of stakeholders should be articulated rather than clustered together in a supposedly homogeneous desirable future.

In addressing the case of the Nanopil, I have elicited the normative positions of different stakeholders by interviewing them and exploring their perspective around a certain social practice and the way in which the emerging technology is expected

---

[12] The Vision Assessment approach was introduced in Sect. 2.3.

to change it.[13] The promises of cost-efficient, comfortable and painless screening attached to the Nanopil suggest a common social problem and a shared normative framework among some actors. Different stakeholders, however, have diverse, and sometimes *diverging*, views of the problem that the Nanopil should address. The same diversity emerges in their judgment of the potential value and drawbacks that the Nanopil might have. For example, when asked to define the problem addressed by the Nanopil, its developers respond that colorectal cancer is usually discovered too late, with the implication of low chances of successful intervention. Their solution is to build a tool for early diagnostics of colorectal cancer in order to enable early surgical interventions with a higher chance of success. For this reason, the Nanopil is presented by developers as a screening device that smoothens a cumbersome, time – consuming and expensive diagnostic practice based on colonoscopy. This novel technology is presented as a more reliable test that can detect the subject's susceptibility of developing colorectal cancer. Furthermore, as the developers explain, the Nanopil is a more user-friendly alternative that does not require people to collect a sample of their own stool to be sent to the lab. Implicit in this perspective is the view that part of the problem of late surgical intervention rests on the unwillingness of people to collect their stool samples for early screening. However, from my interviews[14] it emerged that the general practitioner (GP) and the patient organization identify different types of problems that need to be addressed in the practice around colorectal cancer. According to these actors, the reason for late intervention in colorectal cancer diagnosis stems from the lack of awareness of the Dutch population regarding the risks and symptoms of colorectal cancer. From the patient organization's perspective, a desirable solution to this problem involves an awareness campaign on colorectal cancer. Differences in the problem definition among stakeholders entail differences in the way in which the problem is addressed. The desirability of the Nanopil as a solution to problems in the diagnostics of colorectal cancer is less straightforward when different actors elaborate on what they think to be the cause of problem(s) with a certain practice.

This diversity of visions appears also when stakeholders are asked to identify possible future concerns around the Nanopil.[15] For example, the gastroenterologist imagines a limited use of the pill, because it does not enable a discriminative diagnosis; such a diagnosis is necessary when a symptomatic patient presents him/herself. The GP instead is mainly concerned about the fact that the Nanopil allows users to test themselves in the private space of their home without the assistance of a care practitioner. The patient organization fears that people's privacy is endangered when the result of the test is sent to the user's or her doctor's mobile device. These visions of the concerns around the Nanopil relate to different values. They also depend on the stakeholders' position and epistemological and practical access

---

[13] The design of my interviews is described in Chap. 4.

[14] For an account of how the interviewed experts and stakeholders conceive the preference and needs of potential Nanopil users, see Sect. 4.4.3.

[15] Ibidem.

to the colorectal cancer diagnostic practice. When collecting the visions of different stakeholders on an emerging technology, we come across a richer variety of views on what is valuable or not valuable.

Furthermore, we may come across conflicting judgements on the value of a technology. For example, the developers express their enthusiasm for the possibility of communicating the result of the sample analysis from the pill to an external device via radio-signal because this makes the development of the device easier and speeds up the whole system. However, an advocate from the patient organization warns that such a system poses a possible threat to the patient's privacy. Even though engineers explain that this concern is irrational and unfounded because the system can be secured, for the patient organization, the privacy of the patient needs to be safeguarded even at the expense of the efficiency of the system. Matters of importance for patient advocates differ from those of engineers. For the former, privacy is more important than efficiency. This conflict among values at stake in visions of the Nanopil emerges when comparing expectations of different stakeholders. Explicit or implicit conflicts of values about the desirability of a specific emerging technology can be traced at an early stage in the expectations of current and future stakeholders

In contrast with the general claims of the Nanopil promoters' about desirability, the visions of different stakeholders on the technology at stake are much more diverse and sometimes in conflict. In order to assess the desirability of emerging technologies, divergences in visions need to be made explicit and discussed when engaging in an ethical technology assessment of an emerging technology. Instead of taking the general claim about the desirability of the practice introduced by the Nanopil, I have emphasized diverging problem definitions, concerns and values of different social actors. The assessment of the plausibility of expectations that I propose here broadens the space for alternative heterogeneous views, rather than privileging (even if in a critical way) a dominant, supposedly homogeneous perspective.

5.4 Impacts of Technologies and the Moral Landscape

Thus far, I have argued that assessing the plausibility of expectations concerning the desirability of emerging technologies implies disentangling their unifying discourse. In this way, general claims of desirable futures can be contextualized and specified, while additional aspects and moral connotations are highlighted. The previous sections have drawn attention to how such disentanglement can be carried out by differentiating the variety of moral connotations muddled in expectations on the value of the technology, and by articulating the plurality of normative visions in claims of apparently uncontested desirability. In this section, I contend that expectations concerning emerging technologies need to be assessed also with respect to the expected linear instrumental relation between the technology and its desirable consequences.[16]

An example from history of technology will clarify this point. In the context of an analysis of historical lessons on scenarios of technological futures, Geels and Smits (2000) examine expectations about the social consequences of ICT tools on teleworking. The two authors explain that there is a gap between the expected and the actual consequences of these technologies: it was only when society was confronted with practical problems involving real users and real contexts that the speculative futures proved not to work overly well.

> For instance, employees that are willing to try tele-working at home find that their houses lack space to set up an ergonomically acceptable workplace. Employees discover that they miss the informal and social interactions with their colleagues. The fading distinction between work and private life results in psychological problems in the family. Employees feel that tele-working reduces their career opportunities, as they have less contact with their superiors. On the other hand, managers feel that they have less control over their employees, as the latter are working out of sight. (Geels and Smits 2000: 875–876)

The authors conclude that teleworking is not simply a replacement of an existing practice of localized working. On the contrary, teleworking introduces new practices that re-adjust and change the old practices. The reality of the "social-embedding" of teleworking is more nuanced than the way in which it is positioned when the expectations were emerging. Teleworking does not simply enable employees to work from home or simply reduce costs for employers. Multiple elements have an impact on the success of teleworking, e.g. the aspects employees "miss" of the previous practice; the way in which they conceptualize the difference between private and working life; and managers' feelings. However, this impact of ICT on working practices was not taken into account in the early days. Expectations concerning teleworking were flawed by a linear and instrumental conception according to which technology is simply a means to a (desirable) end.

The discourses on the desirability of the Nanopil articulated in Sect. 5.1 suggest that this device offers a solution to the problem of effective and comfortable/clean monitoring. However, as in the case of ICT, one can expect that the Nanopil will create *new* meanings and practices. Therefore, in order to assess the plausibility of expectations of desirability of emerging technologies, it is important to take into account how the technology reshapes and is reshaped by different areas of life. The notion of "mediation", already introduced in Chap. 3, provides a conceptual tool to address the question of the plausibility of the expected relations between emerging technologies and desirable consequences.

---

[16] Some parts of this section have been published in Lucivero and Dalibert 2013.

5.4.1 Mediation

In order to explain how artifacts contribute in constructing human epistemologies and practices, Peter-Paul Verbeek explores the concept of "mediation" within the post-phenomenological tradition, drawing on post-phenomenologist Don Ihde (Verbeek 2005). Technologies "mediate" our relationship with the world in two respects. First, let us consider imaging instruments introduced in the obstetrician's room. By allowing parents to see the fetus with greater precision, these technologies alter the way the world is present for future parents. The images of the fetus invite parents to perceive, experience and understand the world in a different way than a non-technologically-mediated perception would allow. This is a form of "hermeneutic" mediation. Indeed, new ways of perceiving and experiencing involve "opening" up the world in a different way and changing the universe of meaning; for example, by being able to see the fetus and its human resemblance, prospective patients may attach a new meaning to it. Re-articulation of meanings and interpretations of reality also affect the values attached to aspects of that reality. For example, being able to see the human figure in a fetus might influence the importance that parents attribute to it or the moral status that they ascribe to it.

Building on script theory,[17] Verbeek emphasizes a second type of technological "mediation" that he refers to as "pragmatic". The technology has a program of action inscribed in it. To use the "script theory" vocabulary, "technical objects define a framework of action together with the actors and the space in which they are supposed to act" (Akrich 1992: 208). In this space, roles and responsibilities are allocated to actors in a way that re-designs the previous practice. The moral connotation of this relation emerges in the delegation of moral actions to the technology. The alarm system integrated in modern cars is activated when the seat belt is not buckled. The artifact's design mediates human actions in the world in such a way that some actions will be allowed and others forbidden. In this sense, the artifact prescribes, obliges, permits, prohibits and disciplines users' behavior. This is what Akrich defines as the "moral" content of objects (ibidem: 219).

The expectation of the Nanopil as a tool to improve screening of colorectal cancer reflects an instrumentalist view that assumes a linear relation between a technology and its consequences. Assessing the plausibility of these expectations implies bringing attention to this instrumentalist misconception and emphasizing the expected mediating character of emerging technologies. In order to do this, this section looks closely at the existing practice that the new technology is expected to "improve" and analyzes how responsibilities and tasks are distributed and how knowledge is gained and interpreted. Then, the fictive scripts of the Nanopil developed in the preceding analyses (see Chap.4) are discussed together with the kinds of responsibilities, tasks and epistemic practices that are expected to be inscribed in this device.

---

[17] Script theory has been already introduced in Chaps. 2 and 4.

5.4.1.1 The Practice of Screening

The importance of monitoring oneself is not an emerging practice: to the contrary, it is quite rooted in our society together with the idea that our body manifests some signs that inform us about our health condition. My grandmother gets worried when my cheeks look rather pale or there are white stains on my nails as she knows that these are signs of poor health. My grandmother also knows that if there is blood in her stool matter, there is something wrong going on in her body and she should contact the doctor. The practice of observing abnormal signs appearing on our body involves noticing something that should not be present. This practice can involve routine self-checking and relates to some feeling of repugnance on the realization of signs of decay on our body: we see pimples, blood, cuts, leakages or crusts, we sense bumps or nodes or we feel pain or tingling.

This routine self-checking differs however from systematic and scientifically informed self-monitoring. When women are instructed to palpate their breasts as routine self-monitoring for breast cancer, they are taught how to *look for* eventual nodes. Nodes do not appear on the body; rather women are asked to search for indications that something might be wrong with their health. The presence of blood in the stool, a change in bowel habits, diarrhea, constipation or a feeling that the bowel does not empty completely, abdominal discomfort, smaller stools than usual, and constant fatigue are symptoms of colorectal cancer[18] which a GP might ask patients to investigate in a daily practice of self-monitoring.

Tests like the Fecal Occult Blood Test are similar to breast palpation in the sense that the users are asked to interact with their body (or with a product of it). However, these tests differ in one important respect: while the subject of breast palpation can experience the problem herself by sensing a node under her fingertips, the subject of the FOBT does not have direct experience of the problem. Her interaction with her body (or its product) ends with the act of collecting the sample. Subsequently, the responsibility of monitoring is transferred to the lab and eventually to the GP who communicates the result. In this practice of monitoring, the subject is detached from the experience of her health condition.

5.4.1.2 The Nanopil: Allocating Actions and Responsibilities

The Nanopil is a form of "mediation" between the screening subjects and their experience of their health condition. As I explain in the following section, this mediation is both "pragmatic" and "hermeneutic". First, the Nanopil mediates the actions of the monitoring subject by discouraging or inhibiting actions in which the subject is asked to pay attention to her own health condition. The discourses about the Nanopil propose a "comfort trend", emphasizing the desirability of a test that is acceptable, easy and patient-friendly. By being able to test oneself in the comfort of one's own home, whenever one wants, and by freeing the user from being

---

[18] See http://www.testsymptomsathome.com/mtl01_colon_facts.asp.

5.4 Impacts of Technologies and the Moral Landscape

dependent on laboratories for results, the Nanopil is expected to fulfill this promise. Furthermore, this device is presented as a clean modern test that saves the user from the unpleasant task of sampling her feces.

These ideas are inscribed in the Nanopil's design. The miniaturization of the analyzing platform and its integration into a capsule allows the user to ingest it. The manual collection of samples becomes superfluous, since the pill gathers the sample and analyzes it from within. In this sense, the pill takes care of the whole monitoring process. The screenees are left with information on their mobile phones rather than having to involve themselves in an active and unpleasant practice. The screenees do not have to move and touch their body as in breast or testicular cancer self-screening; they are relieved from the task of peering at their skin to map new and abnormal moles; and they do not have to bend over the toilet to collect feces samples. The technology is expected to liberate people from the discomfort of monitoring, the distaste of dealing with their body, and the embarrassment of describing repugnant signs and symptoms to their GP. The pill liberates users from this awkward link with their possibly diseased body.

However, the Nanopil can also be expected to allocate some tasks to the users. This practice of self-monitoring requires the screenee to perform some tasks. In contrast to the FOBT, the Nanopil does not require the user to interact with her stool matter, to sample it and send it to the lab. However, the user is expected to perform other tasks, like ingesting a laxative before taking the pill. This task is inscribed in an artifact, since one of the main conditions for the pill to work is the ingestion of a laxative to clear up the bowel and to allow the pill to traverse it. Moreover, depending on the interface chosen to communicate results to the user, the user will be required to either put on a belt and receive a text on her mobile phone or look at the color of the stool. Such a test requires strong self-discipline and clashes with some standards of wellbeing and user-friendliness that the user might have.

The user will be asked to perform some tasks, but this work remains currently "invisible" (Star and Susan 1991; Oudshoorn 2008) in the expectations of the Nanopil. The NP does not simply improve a current practice, but creates a new practice.[19] Within this new practice, responsibilities are re-distributed among actors and technologies. For example, adequate performance of the preparatory tasks prior to ingestion of the capsule becomes the user's responsibility rather than the responsibility of the medical personnel or the device itself. The adequacy of the sample collection is a responsibility of the pill (or its manufacturer).

5.4.1.3 The Nanopil: Changing Meanings and Epistemic Responsibility

The NP also "mediates" in the same way a thermometer would do. Reading off the pill is like reading off a thermometer in the sense that the device tells something about ourselves without resulting in a direct sensation. The idea that the pill will be

---

[19] See also Annemarie Mol's analysis of the role of the blood sugar measurer in changing self-monitoring practices of diabetes patient (Mol 2000).

better than other available screening devices (excluding the colonoscopy) is grounded on the promise of molecular diagnostics. Recent trends in molecular biology support the view that self-monitoring, such as "peering into the toilet", is not enough to detect early disease stages: there are some phenomena that cannot be observed by the naked-eye. A currently available screening device like the Fecal Occult Blood Test (FOBT) detects the presence of blood in the feces that is hidden ("occult") to human beings, but visible when a sample of stool matter is analyzed in the lab. The Nanopil brings this observation to a new level of molecular investigation. By detecting molecular markers in the intestinal liquid, the Nanopil seeks a different type of "sign" than the FOBT does. The latter detects the presence of blood in the feces. This could be interpreted as a sign of the presence of a tumor that causes the intestinal walls to bleed. The FOBT provides information about a disease in a stage that might be already advanced. Furthermore, the presence of occult blood in the stool could also be a sign of something else, for example the inflammation of anal veins (hemorrhoids). Finally, the absence of blood does not necessarily indicate the absence of a tumor: indeed, the tumor might be growing but not bleeding. The Nanopil provides information that differs from that of the FOBT; it provides information about a cancer that does not yet exist, but has the molecular triggering conditions that can lead to its development. In fact, the pill detects an abnormal status before any (visible or occult) symptom occurs. By analyzing the molecular mechanisms that underlie the disease, the pill enables detection of the disease at a much earlier stage, when it is still invisible. In this way, a therapeutic or surgical intervention can take place at an even earlier stage, increasing the chances of survival and reducing health care costs.

Molecular knowledge is considered to be superior, because it is more accurate than the behavioral knowledge; it offers a means of returning to the subcellular, molecular level,[20] a level that is expected to be more informative. Our visible body is less informative than our invisible cells according to molecular medicine. It contains less information about ourselves, or it gives us information at a stage at which we cannot intervene with the same efficiency. "The pill knows you best", better than you even know yourself.

Thus, on the one hand, the pill is presented as desirable within a "monitoring" discourse in which health monitoring is portrayed as a moral responsibility towards ourselves and society at large. On the other hand, trust in the pill builds on a molecular trend that indirectly implies the incompetence of the user to effectively monitor her body. The technology is presented as a more efficient way of self-monitoring that transcends our physical body; in this way, while still burdened by some practical responsibility towards ourselves, we are relieved of what we can refer to as "epistemic responsibility" (Code 1987). We are not responsible for the resulting information regarding our health condition because the device does not facilitate the collection, information, processing, and understanding of information.

---

[20] A similar remark is made by Nordmann (2007a) on the assumptions behind the idea of efficiency of nanomedicine.

The expectation that the Nanopil will make screening practices more reliable does not take these aspects into account. The Nanopil can be expected to contribute to a change in the way we self-monitor our health, in addition to the way in which we relate to our body. It has an impact on our practices of being ill, being healthy, and being concerned about our health.

5.4.2 The Co-production of Technology and Morality

By altering meanings and prescribing behaviors, the Nanopil can be expected to redistribute responsibilities among actors in a social practice. The Nanopil, however, is not a direct cause of these changes, instead it is part of and reinforces ongoing "trends". Along with other technologies, the public discourse on health and early diagnosis, social infrastructures and current morals, Nanopil helps to sustain and expand, for example, trends towards monitoring, comfort, and molecularization. In this sense, new and emerging technologies interact with morality in a symmetrical way: current morality and values justify expectations on emerging technologies and are inscribed in their design, but technologies also change this morality. They "co-produce" one another (Jasanoff 2004).

Technologically induced moral change is more than adaptation to the new technology; it reflects deep changes in individuals' and societies' network of values, perceptions, concepts, standards, norms, and habits (Swierstra 2010 and Swierstra et al. 2009). This occurs because the innovative force of new technologies creates *problematic situations* for which new normative solutions are required (Keulartz et al. 2002). Take for example, organ transplantation technology.

> The development of transplantation technology received important support from the optimistic belief that technological progress is an important moral value in itself. Furthermore, as soon as the technological opportunity to help people appeared on the horizon, it created the moral obligation to further pursue this technological trajectory. [...] From the moment this crucial technological innovation made it possible to help patients, the corresponding moral obligation was quickly established. But this techno-moral obligation raised new moral concerns. (Swierstra et al. 2010)

On the one hand, morality is modified to adapt to the new situation, for example, the concept of "death" needs to be redefined with the development of technology for organ transplantation. On the other hand, the technology is modified and technological solutions are sought and proposed to address the moral concerns that other technologies have created. For example, when transplantation technologies created the possibility of saving many people's lives, they also created the moral concern of extracting living organs from a donor. The concept of "brain death", together with new devices to measure the brain activity, offered a solution for establishing whether a person is dead and whether it is possible to proceed with organ transplantation. However, these solutions raised new moral problems with respect to the scarcity of organs and the need for criteria for their fair distribution.

The desirability of the Nanopil is legitimized by the mobilization of values that are supposed to be shared, and non-controversial in current Dutch society: autonomy, comfort or affordability. However, as explained above, the expectations of the Nanopil assume that people can monitor their body and look for signs (markers) of disease before an individual is aware of them. Boenink (2010) has observed that molecular medicine introduces new epistemologies that reformulate the traditional concepts of "health" and "disease". New epistemologies also lead to change in normative definitions: for example the meaning of what counts as "disease" is central in decision of what is considered as treatment and therefore paid for by private or publich health insurance.

The expectations that the Nanopil will improve the current state of affairs in the diagnostic practice of colorectal cancer underestimate an important aspect: the meaning of the values against which the desirability of this technology is *now* defined will change over time, *also* because of the Nanopil itself. Technology and morality interact at a deeper level than techno-scientists seem to expect. Assessing the plausibility of expectations on how emerging technology will achieve some desirable social goals implies a critical revision of the linear instrumental assumption on which they rest. New technologies will not have *one* desirable consequence. They will be part of more complex practices in which epistemologies, moralities and normativity will be adjusted.

5.5 Conclusion

This chapter has shown that general claims regarding the desirability of an emerging technology often draw on a superficial unifying rhetoric of supposedly shared values. These discourses on the desirability of the Nanopil implicitly or explicitly refer to some values (autonomy, care, comfort, efficiency) that the technology is supposed to promote. The Nanopil is said to offer "a solution for a social problem", "addresses a need" and "improves our current clinical practice". The analysis conducted in the chapter demonstrates the ambivalence and contradictions inherent in these expectations. It did so by highlighting and bringing forward, the moral values and ethical controversies that would remain otherwise hidden in the unifying promises around the Nanopil: the moral connotation of possible design choices, the latent ethical controversies in stakeholders' normative positions, the potential technology-mediated changes in the current moral landscape and value framework.

As argued in Part I, emerging technologies are available for ethical assessment in the form of expectations, as projections of potential futures. Such expectations provide unstable ground for ethical reflection as they often play a strategic role in the innovation process, functioning as guiding visions, which depict a desirable world. Although assessing these visions of desirable worlds is important for democratic deliberation on emerging technologies (see Chap. 1), such normative assessments should be underpinned by an assessment of their *plausibility* (see Chap. 2). How can the plausibility of these visions of technology-induced-desirable-futures be

5.5 Conclusion 119

evaluated? To address this question, the Nanopil has been introduced as a case study in Part II. Here, on the basis of theoretical, methodological and empirical studies – mainly drawn from the disciplinary fields of Science and Technology Studies (especially Actor-Network Theory) and Philosophy of Technology – a three-step strategy has been outlined:

STEP 1. Analyzing expectations of the technical artifact. Through interviewing technology developers, reviewing scientific literature, and observing laboratory practices, I collected information about: the origins of the concept, the arguments given for its feasibility; the uncertainties and challenges faced by researchers; and the different components of the future artifact. This *situating* strategy enabled me to *thicken* the original vague expectations on the emerging technology under observation, i.e. the Nanopil. This analysis allowed me to point out alternative, co-existing designs (i.e. wireless/blue bolus), to rule out some implausible scenarios on the basis of the current state of the art (i.e. the availability of the Nanopil at the grocery store), and to point out some conditions for the artifact to work (i.e. the use of a laxative) (Chap. 3).

STEP 2. Analyzing expectations of the context of use of the emerging technology. Using interviews with a broad range of social actors and a literature review on colorectal cancer diagnostics (clinical practices and policies), I collected information about: the current practice in colorectal cancer diagnosis and screening, the considerations of actors, and the problems they face. In this way, I was able to thicken the descriptions of the context of use. I then asked the interviewed stakeholders to comment on the plausibility of technology developers' scripts, from their situated perspectives. This detour allowed me to dismiss some scenarios as implausible, given current social practices (for instance the blue Bolus pill in a national screening program), and to point out possible resistance and controversies (doctors against point of care devices and users against laxatives) (Chap. 4).

STEP 3. Assessing the plausibility of the visions that an emerging technology will have desirable consequences. Based on the previous analyses of the expectations of the artifact and its use, in this chapter I first showed that coexisting alternative designs are muddled within the expectation of "one" technology and that have *divergent* moral connotations. Then, I pointed out that ideas of what is desirable in a certain social practice is not as homogenous as it appears in the original visions, but it varies according to *different* stakeholders. Finally, I stressed that the expected linear relation between technologies and their desirable consequences neglects the *various*, mutual and complex interactions between technology and morality. This analysis allowed me to emphasize the importance of teasing out heterogeneous scenarios of emerging technologies, rather than clustering them together in vague unifying promises (This chapter).

This three-step analysis is not a prescriptive assessment because it does not offer a normative answer to the question of whether the Nanopil *should* be developed or commercialized or how this *should* be done. It is, however, preliminary to such a prescriptive endeavor. Indeed, it emphasises that before asking the question "is the Nanopil a good technology?" we need to assess what expectation about the Nanopil being "good" is plausible. Such plausibility question does not simply focus on aspects of technical and scientific feasibility; instead moral values are considered as active drivers and actors in technological innovation, rather than just background or rhetorical devices. Such assessment of technological visions draws attention on the values inscribed in design choices, on the normative positions held by potential users or stakeholders as well as on broader cultural values. Assessing the plausibility of visions for an early ethical assessment of emerging technologies means, therefore, to appraise these visions against these enriched scenarios, wherein moral values play a role in the realization, adoption and evaluation of the future technology. My claim, which I will further elaborate in Chap. 8, is that this type of analysis facilitates a grounded normative discussion. Rather than favoring one side or the other in public controversies regarding the desirability of certain types of technological innovations, such an analysis disentangles normative issues in grand visions and in this way creates the conditions for a meaningful and normatively inclined deliberation that takes place among scientists, regulators, users, policy makers or the publics.

The Nanopil has provided an exemplar case of expectations the plausibility of which needs to be assessed. Before reflecting on the role of such analysis for the purposes of ethical technology assessment (see Chaps. 7 and 8), the three-step plausibility assessment so far developed will be applied to another case, in order to further explicate the methodology. Thus, Chap. 6 shows how the methodology of plausibility assessment is applied to another emerging scientific and technological endeavor, i.e. the Immunosignatures.

References

Akrich, M. 1992. The description of technological objects. In *Shaping technology building society: Studies in sociotechnical change*, ed. W. Bijker and J. Law. Cambridge, MA: MIT Press.
Bijker, W.E. 1987. The social construction of Bakelite: Towards a theory of invention. In *The social construction of technological systems: New directions in the sociology & history of technology*, ed. W.E. Bijker, T.P. Hughes, and T.J. Pinch. Cambridge, MA: MIT Press.
Boenink, M. 2010. Molecular medicine and concepts of disease: The ethical value of a conceptual analysis of emerging biomedical technologies. *Medicine, Health Care and Philosophy* 13(1): 11–23.
Code, L. 1987. *Epistemic responsibility*. Hanover: Published for Brown University Press by University Press of New England.
Feenberg, A. 1995. *Alternative modernity: The technical turn in philosophy and social theory*. Berkeley: University of California Press.
Geels, F.W., and W.A. Smits. 2000. Failed technology futures: Pitfalls and lessons from a historical survey. *Futures* 32(9–10): 867–885.
Grin, J., and A. Grunwald. 2000. *Vision assessment: Shaping technology in 21st century society towards a repertoire for technology assessment*. Berlin: Springer.
Hughes, T.P. 1987. The evolution of large technological systems. In *The social construction of technological systems: New directions in the sociology & history of technology*, ed. W.E. Bijker, T.P. Hughes, and T.J. Pinch. Cambridge, MA: MIT Press.
Jasanoff, S. 2004. *States of knowledge: The co-production of science and the social order*. London/New York: Routledge.

References

Keulartz, J., M. Schermer, M. Korthals, and T. Swierstra (eds.). 2002. *Pragmatist ethics for a technological culture*. Deventer: Kluwer Academic Publishers.

Latour, B. 1987. *Science in action: How to follow scientists and engineers through society*. Cambridge, MA: Harvard University Press.

Lucivero, Federica, and Lucie Dalibert. 2013. Should I trust my gut feelings or keep them at a distance? A prospective analysis of point-of-care diagnostics practice. In *Bridging distances in technology and regulation*, ed. Ronald E. Leenes and Eleni Kosta, 151–163. Oisterwijk: Wolf Legal Publishers.

Melchior, M. 2009. Doctoren met nanotechnologie. *Medisch s.: MIT Press*.

Mol, A. 2000. What diagnostic devices do: The case of blood sugar measurement. *Theoretical Medicine and Bioethics* 21(1): 9–22.

Nordmann, A. 2007. Knots and strands: An argument for productive disillusionment. *The Journal of Medicine and Philosophy* 32(3): 217–236.

Oudshoorn, N. 2008. Diagnosis at a distance: The invisible work of patients and healthcare professionals in cardiac telemonitoring technology. *Sociology of Health & Illness* 30(2): 272–288.

Pinch, T.J., and W.E. Bijker. 1984. The social construction of facts and artefacts: Or how the sociology of science and the sociology of technology might benefit each other. *Social Studies of Science* 14(3): 399–441.

Reuzel, R., and G.J. der Wilt. 2000. Technology assessment in the health care area. A matter of uncovering or covering up? In *Vision assessment: Shaping technology in 21st century society towards a repertoire for technology assessment*, ed. J. Grin and A. Grunwald, 53–71. Berlin: Springer.

Star, Susan Leigh. 1991. Invisible work and silenced dialogues in knowledge representation. In *Women, work and computerization*, ed. I. Eriksson, B. Kitchenham and K. Tijdens K, 81–92. Amsterdam: North Holland.

Swierstra, T. 2010. Het huwelijk tussen techniek en moraal. In *Moralicide. Mens, techniek en symbolische orde*. [Jaarboek Civis Mundi i.s.m. Rathenau Instituut], ed. Marli Huijer and M. Smits, 17–35. Rotterdam: Lemniscaat.

Swierstra, T., H. van de Bovenkamp, and M. Trappenburg. 2010. Forging a fit between technology and morality: The Dutch debate on organ transplants. *Technology in Society* 32(1): 55–64.

Swierstra, T., R. van Est, and M. Boenink. 2009. Taking care of the symbolic order. How converging technologies challenge our concepts. *NanoEthics* 3(3): 269–280. Springer Netherlands.

van den Berg, A. 2009. De kunst van het kleine, in Brinskma and van den Berg, De kunst van de wetenschap, Redevoeringen 48ste Dies Natalis, Universiteit Twente.

Verbeek, P. 2005. *What things do: Philosophical reflections on technology, agency, and design*. University Park: Pennsylvania State University Press.

Part III

Chapter 6
Expecting Diagnostics, Diagnosing Expectations. The Plausibility Framework in Use

> *The quest we undertake is as down to earth as it goes below surface. (John Grin 2000: 28)*

Abstract "A world without patients": this is the motto of the Arizona State University spin-off company manufacturing microchips for research on "Immunosignatures". By its developers, immunosignatures (ImSg) are presented as a "technological revolution" able to transform diagnostics and improve the American healthcare system. How is the plausibility of these expectations constructed and how can it be assessed? Chapters 3, 4 and 5 of this book use the Nanopil to justify and explain the single steps of a framework for analyzing the plausibility of expectations of emerging technologies. This chapter illustrates the wider applicability of this framework using the case of ImSg. After an introduction on the promise of this technology and a description of the research design, the two central sections analyze expectations in relation to this technoscientific project and its context of use. Based on these analyses, the plausibility of expectations that the ImSg will bring about desirable outcomes is discussed. The chapter concludes with a reflection on the plausibility assessment framework through a comparison of the cases of the Nanopil and Immunosignature.

Keywords Immunosignatures • Plausibility • Expectations • Laboratory

6.1 Immunosignatures and the Healthcare Revolution

The Center for Innovations in Medicine (CIM) is one of ten research centers within the Biodesign Institute at Arizona State University (ASU) devoted to the development of "purposeful research to solve urgent societal challenges".[1] The Center's mission is to develop

---

[1] An abridged version of this chapter has been published as Lucivero F. The Promises of Emerging Diagnostics: From Scientists' Visions to the Laboratory Bench and Back. In: van der Burg S, Swierstra T, editors. *Ethics on the Laboratory Floor*. Basingstoke: Palgrave Macmillan; 2013. p. 151–67.

innovative research that attempts to transform our understanding of diseases. In many cases, innovation requires that we put aside what we think we know and start fresh.[2]

The CIM co-director confidently asserts that they aim at creating "a world without patients". This is also the motto of the spinoff company that he has founded to manufacture microchips for research on "Immunosignatures". This is how, in a recent video from 2010, the lab director describes this technology for "harnessing the immune system's diagnostic power":

> One of the power project in my Center is to have well people monitoring their health in a comprehensive way so that they can detect early any aberrations, anything that starts to go wrong with their health and they can do it early, and act early. We think that this is probably the most important thing that we do around the health …economics in United States also. […] We realized that we have to develop a system that is cheap, very simple and very comprehensive, so that well people can use it all the time. There is a very powerful aspect of that, because it means that you are always normalizing your health with respect to yourself. Right now in the biomarker world in medicine, as we live right now, we are normalizing our markers to the whole population, generally not to ourselves, because we don't take them frequently enough to do that. So, this was our goal. We were trying to figure out how to do that and we knew that it had to be a *technological revolution* in order to be able to do these kinds of things. We finally came on this really *simple concept* and that was: you have millions, billions of antibodies in you and if those antibodies were always registering your health status and we had to look at that whole repertoire, and get a signature of your antibodies *in a simple way, we might be able to revolutionize diagnostics*. We used those same arrays with peptides that we were using to develop synthetic antibodies, and we said what happens if you put a drop of blood from somebody on there, and you wash it off and detect the antibodies? Well, it turns out that you get this signature; you get 10000 spots lighting up at different levels that basically finger-prints your antibodies. We said, maybe there is a change when something goes wrong. And sure enough we have tested over 20 different diseases now. Everyone shows *its own distinctive signature*. So we can normalize to yourself and when something happens, that signature changes. The beauty of it is that *measuring antibodies is so simple*. We can literally take less than a drop of blood, put it on a little filter paper, send it through the mail, even in Phoenix in the summer, take that filter paper and measure the antibodies on it. The signature is just as good as if we had measured the blood directly. So what we envision now is a health monitoring system where people are regularly sending in […] a little thing of saliva or a drop of blood, goes to a central place, they monitor, that information goes back to people and they can tell what their health status is. And that's the big "we are gonna go" that we are shooting for with this. What we show in this first publication is that this works very well to monitor infectious diseases, although we have other projects on cancer and Alzheimer that are going on.[3]

Immunosignatures are a "technological revolution" able to transform diagnostics and improve the American healthcare system. The desirability of such a system is that it will relieve the national economy of healthcare costs – costs which are unsustainable given that technology, hygiene standards, and healthier lifestyles have prolonged life expectancy and standards of medical care. On the one hand, people live longer and this ageing population requires more medical care; on the other, new biomedical research offers new, increasingly effective, but often expensive, treat-

---

[2] From the Centre for Innovation in Medicine website: http://www.biodesign.asu.edu/research/research-centers/innovations-in-medicine

[3] Video available on http://vimeo.com/12370576. Transcript and *emphasis* are mine.

ments. Frequent personal monitoring is expected to prevent people from becoming chronically sick and thus permanently expensive for society. In this context, ImSg provides a system "to have well people monitoring their health in a comprehensive way and detect early any aberrations, anything that goes wrong with their system, and act early". This is done in a "simple" way: immune system activity can be disclosed by the detection of antibodies. The user will put a drop of blood on a piece of filter paper and send it by mail to a laboratory; the laboratory will analyze it; and the information will then be sent back to the user, offering her knowledge of her personal health condition at a certain moment in time.

The concept behind ImSg has existed at CIM since 2007, when the project was known as "Doc-in-a-box". This doctor in a box was expected to develop as a portable device, standing on the kitchen table and used to monitor people's healthcare status. Although the idea of Immunosignatures originates from the concept of "doc-in-a-box", researchers believe that the kitchen table device is a long-term vision, while "Immunosignatures" will come in the near future. The Center's co-directors founded a spin-off company to support this project and to inquire into possibilities for marketing it as a direct-to-consumer test. Small kits have been assembled with a filter paper, a lancet to prick the finger, and instructions on how to send a biological sample to the CIM in order to have personal Immunosignatures detected.[4] ImSg is an emerging technology and many promises circulate about its feasibility and desirability. In Chap. 2 it was argued that initiating an explorative reflection on the desirability of technology at an early stage of development is important in order to improve a process of democratic deliberation. However, because of the uncertain status of expectations, this reflection has to be anchored to solid epistemological grounds. The framework developed so far provides a methodology to assess the plausibility of these expectations.

Immunosignatures are a particularly interesting case to address because it presents some similarities with the Nanopil. First, although visions about the future of Immunosignatures are raised in the context of an academic research group, they are extremely application-driven. Second, the focus on diagnostics, personalized medicine and self-monitoring provides a common theme with expectations regarding the Nanopil. Furthermore, thanks to the previous engagement of the CIM group in an earlier workshop organized by the CNS-ASU/Consortium for Science, Policy and Outcomes (CSPO),[5] it was possible to track the evolution of these expectations over the last 3 years,[6] from a stirring vision of a "doc-in-a-box" to the current idea of "signatures of health status".

The analysis of expectations of Immunosignatures described in this chapter applies the three-step framework explained, justified and exemplified throughout Chaps. 3, 4 and 5 (and summarized at the end of Chap. 5). This framework unfolds in three steps, each building on the other and consisting of: (1) thickening descriptions of the expected technology; (2) sketching out fictive scripts describing the

---

[4] This kit had only been used for research purposes at the time of my fieldwork.
[5] See Selin 2007 and 2008.
[6] The 3-years timespan refers to the time the study was conducted (2011).

context of use and then asking actors (or experts) working with these contexts to assess their plausibility; (3) disentangling the multiple conflicting values implicit in different technological platforms, the heterogeneity in normative visions across different stakeholders, and the different interactions of the technology-in-use with morality.

6.2 Research Design

The main goal of this case study was to test, in a different context, the plausibility framework developed and exemplified in the case of the Nanopil. To do this, I designed my research methods to assess the plausibility of the case of ImSg by building on the experience gained during fieldwork on the Nanopil. In designing my fieldwork, I had to consider that the two projects were different. For example, Nanopil started with a very well defined project promoted by an oncologist with a specific vision of the practice and context in which the technology should be used. In contrast, ImSg was conceived of by a scientist as a way to "revolutionize health". In this second case, different possible applications and contexts of use were explored by researchers.

In this second case my interviews with technology developers and observations of laboratory practices were also extended, to a longer and more intensive engagement. This allowed an analysis of expectations of the emerging technology which could integrate qualitative interviewing and group interviews with ethnographic observations (Tedlock 2000). I was involved at the CIM from November 2010 until January 2011 as "embedded philosopher".[7] I was provided with a desk, introduced during a lab meeting to the whole group, and had access to all facilities, meetings and activities. I attended laboratory and project meetings twice a week, reviewed and discussed scientific papers and project proposals with researchers, and conducted participant observation during meetings and laboratory activities. I also held semi-structured interviews with several members of the research group: in some cases I had up to five interviews with the same person. Besides attending outreach events in which ImSg was presented to a lay audience, I also participated in informal activities (lunch conversations, the Christmas party). Finally, I presented my research twice at weekly laboratory meetings.

This interaction with researchers was initially aimed at understanding the science and technology of ImSg and their role in the project. When I became more acquainted with the scientific and technological aspects, it became clear that the context of use for the ImSg was more broadly defined than in the case of the Nanopil.

---

[7] The expression "embedded philosopher" is an adaptation from "embedded humanist" that Fisher and Mahajan (2010) use to refer to a 33-month period of engagement of Fisher in Mahajan's Nanotechnology Laboratory. As a matter of fact, my presence at the Biodesign Institute had been kindly sponsored by the Socio-Technical Integration Research (STIR) project, Erik Fisher's brainchild (https://cns.asu.edu/research/stir). See also Fisher (2007) and Schuurbiers and Fisher (2009).

Different applications were explored by researchers. For this reason, I created some additional instruments to engage CIM researchers in exploring their expectations of the context of use of these technologies, organizing two group interviews with researchers. The first group interview engaged eight participants around the topic of "the challenges and promises of immunosignatures". The goal of this discussion was to collect researchers' discourses of the feasibility and challenges of ImSg. The second group interview engaged fourteen participants and focused on "the applications and practices of Immunosignatures out of the lab". The goal of this second discussion was to collect researchers' ideas about potential applications and to invite them to articulate their descriptions of possible contexts of use.

In a second stage, I interviewed actors who were partially acquainted with the ImSg project in order to elicit their fictive scripts on ImSg. I interviewed experts who had participated in a former workshop organized by CNS-ASU/Consortium for Science, Policy and Outcomes (CSPO) in 2007, about the societal aspects of the doc-in-a-box (the original project on which the ImSg idea was grounded). The goal of these interviews was to gain a "helicopter" view of the current American healthcare system, trends in it, and expected challenges and opportunities for ImSg. The interviewees included: the scientific chief director of the Biodesign Institute, a professor of law and ethics specializing in the legal aspects of emerging molecular medicine, an expert in business models acquainted with insurance company perspectives, and the director of Innovation and System Design department in a well-known medical clinic.

I also selected and interviewed actors who collaborated with the CIM, or who were pointed out by researchers as potential future collaborators. The goal was to elicit their interest in and doubts about the project, and their beliefs about its (im) plausibility. Among the interviewees were a biologist doing research on Neuro-Oncology and Neurosurgery in a clinic in Phoenix, who provided the CIM with samples of patients with brain cancer; a doctor who specialized in Valley Fever, who was involved in the ImSg diagnostics; an employee at the C-Path institute (which facilitates innovations in exiting the university and entering the market), who offered insights into regulatory procedures for diagnostic devices; and the president of a diagnostics company, whom CIM researchers expect to collaborate with in the future. Finally, I interviewed the directors of two centers at the Biodesign Institute involved in the "Partnership for Personalized Medicine", a multi-institution effort that includes the translational genomics research institute (TGen) in Phoenix and the Fred Hutchinson Cancer Research Institute in Seattle. One of these directors has a background in business and extensive knowledge of business models for health care and personalized diagnostics. The other director was trained in oncology and is involved in research on personalized diagnostics. These interviews often required preparatory desk research or provided new literature to review.

The rest of this chapter presents an assessment of expectations around Immunosignatures according to the three-step plausibility analysis explained in Part Two of this volume. First, the expectations of the object-Immunosignatures will be discussed. Then the attention will be drawn on expectations regarding their use. These analyses prepare the ground for the third step: assessing the plausibility of the visions that ImSg will bring about a desirable world.

6.3 Immunosignatures: A "Simple" Concept

ImSg are presented as a "technological revolution" that is founded on a "simple concept": our immune system mirrors our health. In fact, our immune system produces antibodies as a response to the presence of foreign bodies in our system. A read-out of the antibodies in a biological sample would thus provide information about the activity of the immune system. This would reveal information about the health condition of the subject providing the sample.[8] ImSg provide this read-out together with an interpretation of it. In the following discussion, the origin and history of this revolutionary "simple concept" within the CIM will be reconstructed, pointing out the reasons for its novelty and feasibility (Sect. 6.3.1). This concept will be then examined within the research practice where co-existing platforms with different expected uses will be pointed out (Sect. 6.3.2). Finally, the conditions necessary for the ImSg concept to work in a context of use will be stressed (Sect. 6.3.3).

6.3.1 Reconstructing the History of the Concept

According to the CIM co-director, the whole project started as a way to support another research project being developed within the center: the cancer vaccine. The researchers needed to find a way to validate the vaccine and demonstrate that something had indeed changed within the organism. They started with the question: "Can we make an array to measure 1000 *things* in the blood?" They thought about measuring 1000 proteins, as biomarkers of health conditions, and they wanted to use antibodies, on an array, which could catch the presence of these proteins. However, during the testing phase they realized that the different antibodies had recognizable binding patterns. Researchers wondered "what happens if we put a mixture of the two antibodies on the array?": they tried it, and could see a mixture of the two patterns on the array. This finding made them believe that they could retrieve this information from antibodies within the body. They tried with human blood on the array, and saw a pattern that they interpreted as a mixture of the patterns of the different antibodies present in the blood. Antibodies thus became the thing in the blood that they wanted to measure, while the proteins (in fact a form of simplified protein, "peptides") became the "catchers" on the array. "This is what started the immunosignature array and we scaled it up to 10 000 peptides (simplified proteins)". They ran samples of the same person before and after a vaccination and saw big changes for the same person. In fact, these results showed that: (1) it is possible to distinguish the signatures of different diseases in samples from the same individual; and

---

[8] "Our basic premise is that the antibody profile from an individual reflects their health status. If this profile can be displayed on a sufficiently complex array, the particular responses to chronic diseases will be apparent" (Stafford et al. 2012).

6.3 Immunosignatures: A "Simple" Concept

(2) the signatures of different individuals are dissimilar. A senior researcher who has been working at CIM for several years explains:

> And the group at that point split up and this became its own project. Before we thought that you had to scan every single protein in your body, but if we find a way to read this out we don't need to know every single protein, but your antibodies contain enough information.

The original idea has evolved in such a way that what is registered in ImSg is not the presence of certain proteins in the blood, but instead the presence of particular antibodies. These antibody patterns are now considered to have meaning for the health condition of the tested subject.

CIM researchers were therefore originally looking for something else: they wanted to use antibodies as biomarker catchers, without considering that antibodies can be even more informative than biomarkers. The history of the Immunosignature idea is described, by researchers, as a "paradigm shift" in their understanding of the relationship between the detection platform and biological information: what was initially the platform (the antibodies) became the information being sought, and what was the information (the proteins) became the detection platform.

This paradigm shift was accompanied by a second novelty: the use of random peptides (chains of amino acids) on the microarray. As the researchers explained, the current paradigm in immunology is that antibodies are antigen-specific; that is, antibody x will bind to specific foreign body, antigen y. According to this paradigm, in order to "catch" the antibodies in a biological sample, specific proteins should be used in the detection platform (a glass microarray). However, this is not the case on the CIM platform. Figure 6.1 in the Appendix of this chapter (first presented by a graduate student during a laboratory meeting) shows randomly assembled chains of amino acids (peptides) co-located on a glass array (step 4). These peptides are *randomly* generated, in the sense that they present some of the "bricks" (amino acids) which make up specific antigens, but that these bricks are erratically arranged. This means that no one specific antigen is on the glass array. When researchers put a drop of human serum on the array, they expect that the antibodies in the serum will bind to the peptides on the glass plate (Fig. 6.1 step 5). By expecting antibodies to bind to random chains of amino acids, researchers at CIM challenge the paradigm in immunology of antibodies as antigen-specific.

This second paradigm shift is recurrent in researcher discourse. When a researcher returned from a conference, he shared with his group the resistance of the scientific community to what he thought was a simple concept: the group's use of "randomly generated peptides". His audience couldn't grasp how, on their array, they had chains of aminoacids randomly generated by a computer, rather than using chains of aminoacids corresponding to a specific pathogen. The innovation the CIM team presents doesn't fit into the current state of the art, and therefore encounters resistance from the scientific community. During the first focus group I organized, a researcher reflects on possible hurdles for ImSg:

> Another big hurdle, leading to why these arrays wouldn't be useful, is that everybody thinks that antibodies see one thing, except maybe a few weird ones that might see multiple targets, but in reality all AB can see multiple targets […] people get ingrained that the pathogen

gets into the body, and antibodies are generated against it, and just it. Prof. [expert in the field] attended one of our meetings and Stephen asked her "what would you say about monoclonal that recognize a lot of targets" and she replies "well it is a very poor monoclonal" […] eventually things will start clicking…I don't know it might be too big of a comparison, but it is like trying to show that the Earth is round and not flat!

According to the current paradigm in immunology, antibodies are antigen-specific. Monoclonal antibodies are considered able to bind to only one specific antigen. Within this paradigm, it is hard to explain how antibodies can bind to randomly generated peptides that do not correspond to any antigen. The only plausible explanation is that it is a malfunctioning antibody ("a poor monoclonal"). The need to overcome the resistance of the scientific community, "ingrained" in the traditional paradigm of immunology, emerges in many conversations with CIM researchers.[9]

6.3.2 Concepts and Components in Research Practice

"Immunosignature" is a concept that has been coined at CIM. It refers to the fact that it is possible to retrieve specific patterns, unique to each individual, by catching the antibodies in a serum sample. In the first focus group one researcher defines "immunosignaturing" as

> a general *method to 'immunosignature'* the antibody repertoire of an individual […] a readout of antibodies' profile on a random space.

According to this definition, an immunosignature is the actual "image" (or "snapshot") of spots on the array obtained when the glass plate, with antibodies on it, is scanned (Appendix, Fig. 6.2). The specificity of ImSg with respect to other immunoassays is that they do not provide information about the presence of specific antigens, but show general patterns of antibodies. As a researcher explained to me:

> With traditional tests, you're only analyzing the immune response to very defined things. The question addressed by IMS is not "is there a particular disease or infection going on?", but "what is going on?"

---

[9] However, sometimes this controversy lurks in the discourse of researchers working at CIM. For example, CIM researchers talk about "real" peptides to refer to known sequences that are recognized as being the target of specific types of antibodies. They call "artificial" the peptides randomly assembled that they use on the immunosignaturing array. "In science nobody believes you if you don't show that actual stuff in the body. Here you show random stuff, artificial, harder to convince scientific society who only believes in real stuff" or "this is more real…random peptides is a sort of artificial". This dichotomy "REAL versus ARTIFICIAL" suggests a hierarchy in the ontology of the researchers. The peptides synthesized in the lab from random amino acid sequences are less valuable, or less trustworthy, than ones traceable in nature. On one hand, researchers think that the ImSg concept does make sense and that the "conceptual hurdle" for the scientific community can be overcome by showing data that support this unconventional view of how the immune system works: hence publishing, showing evidence, producing results, winning grants, etc. On the other hand, by speaking of "real" or "artificial" peptides, the researchers embed in their daily language the skepticism of the scientific community. There is more. This distinction between "real" and "artificial" epitopes also discloses a difference in what the concept of "immunosignature" means.

6.3 Immunosignatures: A "Simple" Concept

ImSg doesn't provide information about the presence of a specific disease, but about the comprehensive condition of health. Thus ImSg should be a technique, method or platform for identifying immunosignatures concerning a general health situation, rather than a tool for detection of a specific antigen.

However, this apparently clear-cut concept turns out to be much more diversified within research practice.[10] Researchers at CIM do not always use arrays on which random peptides are placed – sometimes they study the antibody patterns specific to particular diseases, and thus obtain an "immunosignature" for a specific disease. For example, a group of researchers were working on a project on biosecurity aimed at testing soldiers for specific bio-threat agents. They had to select specific pathogens (for instance anthrax or smallpox) and place them on the chip. The same procedure was done by a researcher working on an infectious disease typical of Southwestern States (Cocci, or Valley Fever) who had to place specific pathogens onto the array.

Random peptide microarrays are an "unbiased" way to test antibody repertoires, such that researchers do not need to have a "pre-conceived idea" of what peptide (and thereby pathogen) they should use to detect a specific antibody. In this way they can look at "patterns" on the array without needing to know the pathogen. According to them, this is a good way to know whether there is any disease present, rather than pointing out a specific pathogen and how it can be treated. However, in some projects this more specific information is relevant, and therefore a different platform will be used. These considerations are important to this analysis because the concept of "immunosignature" clusters these differences in material practice and application under a homogeneous label. Differentiating amongst these might not change the practice of research, since researchers seem to communicate effectively.[11] However, by analyzing these differences and assumptions in the basic concepts of new science and technology, alternative platforms and applications can be pointed out. For example, ImSg based on a "random space" provides information on comprehensive health status, while chips with specific pathogens on provide information on a specific health condition. As I will show (Sects. 6.4–6.5), distinguishing these technologies and applications is an important step in assessing the plausibility of expectations in ImSg.

---

[10] "Immunosignature" can here be considered as a "boundary object" (Star and Griesemer 1989) plastic enough to adapt to different needs and yet robust enough to maintain a common identity, and allowing negotiation and co-operation of actors around them.

[11] However, after I shared these considerations with the CIM staff during a laboratory meeting, one of the co-directors admitted that this made him reflect on the direction that the center is taking (for an elaboration of how my analysis can play a role in the development process see Chaps. 7 and 8).

6.3.3 Some Conditions for Immunosignatures to Work

One more aspect of the feasibility of ImSg should be mentioned. This technology is expected to provide information about the health condition of a person at a certain moment in time. As the co-director explains in the video mentioned above, it is truly personalized information, a "finger-print" of antibodies:

> There is a very powerful aspect of that, because it means that you are always normalizing your health with respect to yourself. Right now in the biomarker world in medicine, as we live right now, we are normalizing our markers to the whole population, generally not to ourselves, because we don't take them frequently enough to do that. So, this was our goal. [...] Everyone shows *its own distinctivesignature*. So we can normalize to yourself and when something happens, that signature changes. (*My emphasis*)

How do correlations among signatures appear? According to researchers, the similarity between today's or yesterday's patterns is shown by analysis of data collected on the glass plate: samples are clustered in graphs according to similarity between patterns, which depend on the intensity of a peptide – as shown by its illumination within the array – in a certain position (Appendix, Fig. 6.3). Samples of individuals with the same disease are expected to show a similar "pattern". Differences between people with a disease, and those who are healthy, show that there is a pattern of normality and a pattern of aberration. Differences in analyses of different individuals' samples show that there is a personalized pattern, or individual signature.

However, it seems that researchers are not primarily concerned about individual signatures, but rather with standard signatures. During my conversations with researchers, the importance of having a baseline to describe when a signature is normal, when it is abnormal, and when it is going towards a pattern indicating cancer or Alzheimers often emerged. In order to make these differences legible, researchers need to gather enough information about the baseline of normality. In fact, signatures are consistent for the same healthy individual but are very different among different individuals. In the case of a disease, the signatures of more individuals change, and they have a similar configuration, for instance in terms of a "common signature of influenza". Identifying such configurations requires huge statistical effort, given that researchers are looking at differences both between people and between diseases. As one graduate student explained at a laboratory meeting, they need to identify a "standard normal signature", that is

> a reference line of Immunosignaturing, a range which gives an idea how normal individuals (free from any common chronic disease, irrespective of gender and age) respond to a particular peptide on an average. This line has the potential of contributing to baseline any class and also filter peptides which can act as controls, since their overall behaviors is known: a giant step towards 'usage' chips.

The final goal is to differentiate individual signatures, but in order to do so they have to find some common patterns. So "personalized signatures" might be a long-term goal, while standard normal signatures are the short-term goal most researchers are focusing on in their daily activity.

6.4 The Expected Context of Use 135

In conclusion, the analysis of expectations about ImSg science and technology has first of all articulated the novelty of the concept of "immunosignatures". By using antibodies as a source of information about condition of health, and randomly assembled peptides to catch them, ImSg challenges two "paradigms" in the microarray and immunology world. For this reason, according to researchers, it triggers the disbelief of their peers. This analysis has also disclosed different ways that ImSg is done in research practice, and suggested that alternative technologies, with different expected uses, co-exist behind the idea of ImSg. Finally, moving beyond the rhetoric of the promise of a "personalized monitoring system normalized with respect to yourself", I highlighted a condition for the success of ImSg, namely the definition of a "standardized normal signature". When situated in the actual scientific practice of "doing immunosignatures", the promise of ImSg acquires new dimensions.

6.4 The Expected Context of Use

Researchers claim that ImSg is inspired by societal needs and purposes. Expectations around its social embedding circulate during CIM meetings and group reunions. Some of these expectations of use are based on the vision that originally triggered the project as a whole: the "doc-in-the-box"(DiB). DiB was conceived as a portable box with a reader, which would sit on the breakfast table as part of a family morning routine. It was expected to analyze small biological samples (such as a drop of blood or sputum), compare them with the history of the personalized signature and other biological information from the subject, and show an immediate result on its reader.[12] According to the lab director and CIM researchers, ImSg is a more scientifically mature version of DiB, which remains the long-term vision for and ultimate end point of their research. Like DiB, ImSg will provide comprehensive monitoring for healthy people. However, closer analysis of research practice shows that expectations of the use of ImSg are more divergent than these visions suggest. Not only is the "fictive script"[13] of ImSg different from the DiB one, but researchers also use more than one fictive script (Sect. 6.4.1). When assessed through the situated

---

[12] This is the CIM co-director's vision of Doc-in-the-box in 2007, compiled for a workshop on the future social and ethical implications of personalized medicine. "Vision: It is Monday. After taking my HEPs (health enhancement pills) I put the lavager to my nose. It takes a painless wash of the nasal cavity and captures it in a module that is put into the BioSignaturer on the corner of my breakfast table. As I am fixing breakfast the screen on the BioSignaturer comes up. It notes that there are 19 significant variants from my signature last Monday. Looking at my history of signatures and integrating my genotype it warns that I am in the first few hours of an adenovirus infection. This information has also been relayed to my HIM (health information manager), Edith. Edith emails in a few minutes to recommend a course of Zn+Adeno inhibitor that will arrive at my workplace in two hours" (in Selin 2007).

[13] For an explanation of the concept of "fictive script" see Chap. 4.

perspective of other actors, some of these contexts (and scripts) are ruled out, while other possible contexts of use are pointed out (Sect. 6.4.2).

6.4.1 The Many Applications of Immunosignatures

Despite the claimed overlap of the two visions, expectations of the context of use of ImSg and DiB differ in many respects. The idea of using detection antibodies solved the technical challenges of building a box that could analyze data on the kitchen table. In fact, antibodies (the target of the analysis) appear to be very stable molecules which are not easily affected by the environment and do not have to be analyzed immediately. That's why, in 2010, the center's director specified that the samples can be sent by mail "even in Phoenix in the summer". Replacing the idea of the "box" with a shipment system affects the context in which this technology is used. In fact, DiB was expected to be used for DAILY monitoring of ANY kind of health status, while ImSg can be done only REGULARLY because it requires the user to do more work (shipping the sample). The time frame assumed by researchers is "once a week/month". Furthermore, whereas the DiB was expected to detect infections before the symptoms would appear, the current vision of ImSg does not portray it as suitable for detecting infections. Indeed, the researchers note that a typical infection lasts for one week, and that the time between the test, the shipment of the sample, and the result arriving is also likely to be one week. By that time, symptoms may already have been expressed, and the result of the ImSg is useless. In this sense the usability of ImSg is circumscribed to pre-symptomatic monitoring for chronic and long-lasting diseases. Thus while the visions of ImSg and DiB become assimilated, one into the other, important differences are kept out of sight.

Other differing visions of the context of use can be found in CIM researchers' expectations of ImSg. ImSg is presented by researchers as a revolutionary technology and a "disruptive" innovation rather than an "incremental" solution to existing problems. According to this view, "incremental" solutions to a problem are ones that invest in newer, more effective and cheaper treatments for diseases. In contrast, a "disruptive" approach offers a revolutionary solution to problems in healthcare by addressing the stage when people are still healthy, and preventing them from becoming chronically ill. ImSg is not a diagnostic test for telling whether people *are* sick, but a system for pre-symptomatic monitoring of healthy people that tells you whether people are *in the process* of becoming sick. This system is thus expected to enable therapeutic interventions at a stage when no symptoms have emerged.

As such, the CIM directors and one graduate student are exploring possibilities for introducing ImSg onto the market as an online direct-to-consumer (DtC) test. They have been exploring the business models of successful companies, such as 23andMe, that provide DtC genetic tests.[14] *In doing this, they have to address the*

---

[14] See the website where this direct to consumer genetic profile is sold: https://www.23andme.com/

problem of the regulatory constraints that the FDA[15] has placed on online sales of diagnostic genetic tests.[16] Companies like 23andMe have found creative solutions to these constraints, including offering these kits for "personal genetic information" rather than "diagnostics". One possibility for marketing ImSg at this early stage would be to follow this model and offer "raw data" to consumers about their immune system activity. The users would receive information about their immune system activity, without any interpretation of their meaning.

While some researchers address these kinds of questions, most CIM activities seem to focus on defining the "patterns" for specific diseases: infectious diseases, influenza, diabetes, breast cancer, Alzheimers disease. Focusing on pathogens and disease patterns does not necessarily mean that ImSg is expected to be a diagnostic tool for that disease. For example, when researchers talk about the influenza virus, they may be more interested in its value as a discovery tool and model because:

> influenza is a great model system, that only has 8 proteins and it is easily being associated with the patterns that we see on the random array to what the actual sequences are.

However, in some cases ongoing research at CIM is overtly focused on specific diseases. One graduate student collaborates with a medical doctor and expert on Valley Fever. She explained to me how the doctor has provided samples from patients who have been checked for the fungus that causes the disease. Her project aims at exploring the possibility of early diagnosis of this infectious disease; preliminary results in this are promising. Within research practice at CIM, several applications for ImSg are being explored. The most successful current projects at CIM actually focus on its application as a diagnostic tool for specific diseases: ImSg is therefore not only a "disruptive" innovation, but also provides a platform for more "incremental" applications. The unifying vision of a system for monitoring asymptomatic people's general health doesn't do justice to this variety of research.

In the second focus group I organized (see Sect. 6.2), I explored this variety of potential applications. I asked participants to list and describe short-term applications that they envisioned for ImSg. After some discussion amongst themselves, they agreed that it is more likely that in the short term the ImSg platform will be applied to enhance current diagnostic practices for specific chronic and infectious diseases, rather than being a tool to comprehensively monitor asymptomatic people's health status. For example, they suggested that they can identify specific patterns for the Valley Fever infection, which might offer more accurate, cheaper and easier diagnostic tests than are currently available. Other short-term applications envisioned by researchers were: use within vaccine trials; as a diagnosis for cancer, autoimmune diseases, infectious diseases, and chronic diseases; clinical monitoring of cancer recurrence and effectiveness of vaccination; and tools to assess if a certain population (e.g. soldiers) have been exposed to a particular agent.

---

[15] *Food and Drug Administration, the American agency* responsible for protecting and promoting public health through the regulation and supervision of, among other things, pharmaceutical drugs, medical devices and vaccines. See http://www.fda.gov/

[16] See Little 2006.

One of the "fictive" descriptions articulated during the focus group focused on the ImSg as a platform for the diagnostics of a specific chronic or infectious disease. In the case of Valley Fever, researchers drew parallels between this potential system and the current US procedure for testing for strep throat. A symptomatic patient will be asked by their medical practitioner to perform the test through the collection of a drop of blood. The test will then have to be sent to the laboratory for analysis, and, in the case of positive results, the patient needs to go back to the primary care doctor in order to receive treatment.

A second fictive description elaborates a general vision of ImSg as a tool for the comprehensive monitoring of healthy people. Asymptomatic, healthy people will regularly send a biological sample to a central laboratory and receive their results by e-mail. An online system could be in place, in which people can log in, order the testing kit, and receive it at home. After putting the drop on the filter paper, the sample will be shipped, probably in a post office, so that an officer can write down the content and the reason that they are sending a biological sample. A central laboratory can analyze the sample and upload the result onto the online platform, so that users can have online access to their immunosignatures. This fictive script presents two phases. In the short-term, while research on correlation between immunosignatures and health status is still going on, this system is expected to provide information about your immune response, without providing any interpretation. This information could be openly shared and "people" with computational skills might be able to find some correlations. In the long-term, it could be uploaded on some platform (such as Google Health) where people collect and manage their own personal health records.

These fictive scripts discussed by researchers are interesting because they flesh out different ways of envisioning the future of the healthcare system, and the place of ImSg in it. They illustrate how an apparently unitary expectation of a technology opens up very different contexts, with different kinds of questions. The vision of ImSg as a "comprehensive monitoring system for healthy people" thus competes with a vision of ImSg as a "diagnostic tool" for specific health condition in symptomatic patients. Many more details could be added to these descriptions[17]; however, these fictive scripts offer a starting point for exploring the plausibility of expectations of the use of ImSg as expressed by other social actors and experts.

---

[17] For example, where is IMS done? At a primary care consultation every year? Or by individuals in their homes? Exploring the site assumed in researchers' expectations of ImSg is not a secondary matter because different sites define different contextual features: the actors involved (a doctor, a nurse, the pharmacist or a laboratory employee?), the frequency (once a year at the primary care, whenever symptoms occur, according to a personal routine at home), the communication system for the result (face to face, by snail-mail, by e-mail), information storage (in the patient's virtual or paper-based personal medical record, at the GP's office, in an online database), and the purpose of the test (diagnostics, prevention, education, profiles databank building). In each of these contexts, the network of actors changes, as do their relations. Thus each one of these expectations carries a different "script" of how the ImSg system will look and work.

6.4.2 Assessing and Enriching Fictive Scripts from Situated Perspectives

If researchers believe that ImSg can most easily enter the market as a diagnostic device for specific health conditions, other stakeholders and experts assess the plausibility of this expectation differently. One of my interviewees works at the Critical Path institute, an independent, non-profit organization launched by the FDA in 2004 to play a mediating role between scientists, regulatory agencies and industries in order to facilitate the drug development process.[18] He explains how FDA regulations on biomarker tests require specification of the sensitivity and specificity of each element of the test in order to establish the precise accuracy of the test. To date there is no diagnostics test done using "patterns".

> It can be expected that you will have to convolute that pattern and specify for each component what is the specificity and sensitivity

Since 2006, FDA has introduced a class of assays, IVDMIA ("in vitro diagnostic multivariate index assays"), which need more regulation than other tests (Little 2006). These tests screen for thousands of markers in order to predict the likelihood of a certain disease state, disease progression, or response to a therapy, by using an algorithm that gives different weightings to different markers. The impossibility of the physician interpreting the test without the algorithm makes the FDA wary. ImSg, which will use algorithms to interpret patterns, can be expected to raise similar problems. The expert in legal aspects of emerging biotechnologies that I interviewed explains:

> Would FDA's attempts of strengthening regulations be successful, we might wonder whether any company would invest in such the high costs of regulatory approval for a diagnostic test for a specific infectious disease, say Valley Fever, whose potential return might be too low.

The director of the Center for Personalized Diagnostics at the Biodesign Institute similarly shared their experience training as a medical doctor in oncology:

> if you screen for some proteins, you can say that together all these signatures are signs for that disease [...] people might have similar patterns, so we are going to classify these people as type a or type b. This is very different from when a person enters your office and you want to classify her as type a or type b, but this is different because the person is different from the population you have tested it [...] You know that some people belong to a class or another. But is this test good for a patient? NO! Because what if the value of an individual falls in between? What do you tell to the patient? What you want is having no space between the two curves, otherwise you have too much variation and you don't know what to tell to the patient.

If ImSg are conceived as a "tool for the doctor", then they are expected to address the problems that doctors encounter when they need to diagnose a patient. The test has to provide them with the information they need to make a diagnosis – information

---

[18] See http://www.c-path.org/

which will enable them to say something to a patient. In the current regulatory and clinical climate, patterns and algorithms do not provide a trustworthy source of information for giving a diagnosis. Thus the application of ImSg for specific diagnostics may encounter resistance from clinicians.

What about the plausibility of the second vision of Immunosignatures as a pre-symptomatic monitoring platform? As far as it concerns the online, direct to consumer model, from a legal perspective:

> At the moment, there is an ongoing legal action in which FDA evaluates 23andMe, Inc.'s product as a diagnostic test and expects the company to apply to FDA for market approval. The way this legal act will evolve will seal the fate of other products as IMS-based tests for health and wellness monitoring.

In the case of a direct to consumer test, ImSg should be expected to be a freely purchasable product, since reimbursements from insurance companies are unlikely. This was explained by the respondent from the health care business, who is in contact with US insurance companies. From her perspective, a device for the regular monitoring of healthy people which would recognize if you are incubating a flu virus is not likely to be paid for by insurance companies. She explained that insurance companies "are not much of futurists" when making decisions about which treatments and diagnostics to reimburse: the decision whether to invest in a new technology is based on the current situation, including economic utility and clinical validity. The utility outcome of the use of ImSg for healthy people (and the preventive measures that it would entail) should be clearly comparable with the costs of letting the subject have the disease. If ImSg detects early stages of chronic diseases, insurance companies would likely not reimburse the service for the whole population. The respondent seemed more keen on the prospect of insurance companies being available to cover expenses for high risk groups, such as elderly patients, pregnant women, and people with cardiac risk or diabetes – population groups that are 'expensive' because of their high risk of getting sick. Monthly monitoring can be justified and accepted only if the benefits are clear:

> For example, cancer patients are very expensive for insurance companies; therefore a system that would keep them out of the hospital would save a lot of money to payers and would therefore be welcome. But knowing if I get the flu or not, is not of much interest for them, it doesn't save them enough money to be worth it. It is a lot easier to present prospective studies that show that Immunosignatures can be convenient for a group of high-risk patients than for the population in general.

The "world without patients" motto is too futuristic for "payers" who would not see the long-term benefits and focus instead on short-term gains. The idea that patients are in charge of themselves might also, the respondent explained, encounter resistance:

> Payers have reduced hospital time significantly in the last 20 years. There is less hospital and more outpatient care. Hospitals, cancer centers have contracts (incentives) with the insurances that they have less clinicians and having a better quality. But they always want a clinician to interpret results for the patient. They don't want self-medications. That's the big issue.

6.4 The Expected Context of Use

A condition for ImSg to be taken seriously by payers is that information about the result should pass through a clinician. In this regard, ImSg could be integrated into the new concept of the "medical home" that has recently been incentivized by many payers. This is the idea that primary care physicians (or general practitioners) keep track of the patient's complete medical history across visits to different specialists. Payers are incentivizing general practitioners to set up an information management system that enables them to administrate patients' medical records in an efficient way. In this sense, according to the respondent (expert in designing business models for healthcare), ImSg could

> allow the general practitioner to have more awareness of what is going on with the patient (in a high risk group) and this would be much appreciated by the payers

Based on the perspective of insurance companies, a third vision emerges. In this vision, ImSg is not an enabling tool for the patient's self-care and control, but a system that makes the general practitioner more aware of what is going on with patients in a high-risk group.

This third vision of ImSg as embedded in a broad system, in which self-monitoring functions to improve the quality of the health provider's service: this is also the vision of the director of the Innovation and System Design Department at a major American clinic. He explained that the mission of his department is to integrate predictive and information technologies to manipulate historical data and develop predictive profiles for patients who are about to face a surgical intervention. This would help anticipate what kind of care the patient will need and how long she will be in the hospital. In this context, it is important to have enough information about patients in order to reduce their hospital stay and thereby reduce costs. According to this stakeholder, ImSg might be a helpful tool in retrieving more information about patients or prospective patients and making the system more effective.

This actor explains how his clinic is investing in programs of home care or telemedicine, such as the "Heart Care at Home" program, which has drastically reduced patients' stays in the clinic.

> The distinction between doctor's office and home medical care will blur in the future [...] It has a huge benefit in the healthcare system: the healthcare stops being episodic and becomes continuous.

In this vision of ImSg in a "home medical care" context the technology could produce information that joins a large information stream that the system can analyze. The connection of remote technologies changes the relation between the provider and the consumer: ImSg as a point-of-care test, carried out by patients at home, could be part of this home care system and help to make contact between patients and care providers "less episodic" and "more continuous". This respondent also refers to a trend of care providers helping patients to think about their health future and being proactive about it. In their programs, they integrate traditional medicine with complementary approaches around prevention and education about managing the effects of chronic disease through lifestyle modification (nutrition, exercise and stress management). ImSg could provide a tool for raising awareness of the connection between people's immune response and lifestyle.

Similarly, the Director of the Center for Sustainable Health at Arizona State acknowledged the importance of educating people for improving health business and management. He stressed that a change in citizens' attitude towards their health management is a key condition for improving healthcare at a national and international level. In his view, institutions have to invest in order to modify citizens' ideas about their personal responsibility towards society for keeping themselves in a good condition of health. Educational programs, awareness campaigns, and simple game-based persuasive technologies that enhance people's awareness of or control over their health are possible paths in this direction. This respondent envisions ImSg as a tool for changing citizens' ideas of responsibility by making them more aware of their health conditions.

The same technology, then, can be envisioned by scientists as acting in a non-medical context to promote a "world without patients" and by care payers/providers as being embedded in a more controlled context of use. According to this vision, the decentralization of care provision (which will be relocated into the patient's home) brings about a more effective system, which in fact involves more systematic control. The individual's home becomes an outpost of the doctor's office.

This analysis of expectations of the use of ImSg thickens generic visions. Some scenarios and contexts of use can be ruled out, such as monitoring a developing flu. New scenarios also emerge such as, for example, the use of Immunosignatures as diagnostic tools for specific diseases. Furthermore, the fictive scripts articulated by researchers have been assessed by other experts and stakeholders. This assessment points out some barriers to the social embedding of ImSg based on current social configurations (FDA regulations, the role of insurance companies). This assessment also indicates opportunities and contexts of use for ImSg based on experts' and stakeholders' interpretation of ongoing trends (for example, "medical home" initiatives by insurances and hospitals).

6.5 Immunosignatures and a Desirable World

The previous sections have set the stage for assessment of expectations of the desirability of ImSg. When asked about the value of the doc-in-a-box, one of the CIM co-directors explained that its value was threefold: economic, for treatment, and for empowerment and education.

> Economic: The health care system now is largely focused on taking care of ill people. This is where most of the dollars are spent. We currently spend $2.2 T dollars (19%GNP) on health care and are projected to spend $4 T (25 % GNP) by 2015. This is not sustainable. One way to avert this crisis is to convert to a pre-symptomatic versus post-symptomatic medical system. We need to have a longer lived, better healthed population. Treatment: Early detection of disease would allow more effective use of even current medications. It would also open the opportunity to develop new classes of medications that act at an earlier,

and presumably easier to treat, stage. Empowerment and Education: Each individual could see their own biosignatures and the implications of these signatures on a regular basis. This should empower them to take more responsibility for their own health and stimulate their own scientific education on what this means.

The ImSg inherits these values from the doc-in-a-box. As pointed out in the video, this innovation is desirable for the American economy. A healthcare problem is outlined in financial/economic terms: healthcare costs are unsustainable for the US economy. ImSg addresses this problem because it enables a "cheap", "comprehensive" and "regular" monitoring system for healthy people. In this way "well people can monitor their health in a comprehensive way and detect early any aberration, anything that goes wrong with their system, and act early". Frequent personal monitoring would allow doctors to intervene early in the case of any aberration and thus to prevent people from becoming chronically sick and permanently expensive for society. Together with this economic value, ImSg is also desirable because it is assumed that early detection makes existing treatments more effective and fosters the development of new treatments that address the early stages of disease. ImSg provides a mechanism for earlier, and therefore more effective, treatment and helps achieve better health. Finally, by displaying information about an individual's health parameters, and showing how these parameters are linked to health conditions, ImSg enables individuals to "see" these correlations. This understanding empowers people because it makes them more scientifically literate and therefore better able to control the health of their body. Individuals are thus empowered to take responsibility for their health. In conclusion, IMS is desirable because it has a money-value for society, a health-value for patients, and a control-value (both as knowledge and as autonomy) for individuals.

The analysis below assesses this general expectation by pointing out alternative moral connotations of the different technological platforms and applications of ImSg (Sect. 6.5.1). It also points out how the desirability of the expected outcome can be differently appraised by different stakeholders (Sect. 6.5.2). Finally, it shows how expectations of ImSg neglect the way that this technology mediates our understanding of reality and actions in society (Sect. 6.5.3).

6.5.1 Articulating Moral Connotations in Different Technological Platforms

ImSg is expected to solve the problem of high healthcare costs. However, we can ask *which* ImSg is expected to have this desirable consequence. In fact, as explained in Sect. 6.3.2, research on ImSg is currently going in different directions. For example, researchers "do immunosignatures" in at least two ways: (1) they use an array with 10,000 random peptides; or (2) they use arrays displaying only some specific epitopes of known pathogens. These two ways of doing ImSg are not only different stages of the validation of a scientific hypothesis: they also address different clinical questions. In the first case, ImSg answers the question "what's going on?"; while in the second case, it answers the question "is there a particular infection or disease going on?". Section 6.4.1 explored the plausible contexts of use of these platforms. The first platform is expected to be used in the context of comprehensive presymptomatic health monitoring, where there is no suspicion of a specific syndrome. Some researchers at CIM are also investigating the possibility of following the example of other companies, and offering this comprehensive pre-symptomatic monitoring as an online, direct-to-consumer test. According to the stakeholders I interviewed, this platform is likely to be enrolled into a "medical home" program for high-risk subjects (Sect. 6.4.2). The second ImSg platform, which addresses specific diseases or infections, is expected to be used in a clinical context in which a diagnostic decision has to be made. These different types of ImSg are placed by researchers on a temporal axis: the use of ImSg for specific diagnostic purposes comes before its use in comprehensive monitoring. However, these visions are not different phases of the same technology. There are, in fact, three expected technologies: (1) do-it yourself (DIY) health monitoring; (2) disease diagnostics; (3) medical home monitoring. Immunosignaturing platforms are therefore diverse both in their techno-scientific components and in the clinical (or more generally, social) context in which they are expected to operate. These emerging "technical codes" (Feenberg 1995)[19] carry different moral meanings.

This is evident if we analyze and compare what each one of them does. The DIY online comprehensive monitoring system is expected to provide online customers with information about their immune system activity. In the short-term vision, this information is provided to clients but no interpretation of its meaning is given (see Sect. 6.4.1). Based on the business model used by other direct-to-consumer test providers, it is assumed that individuals own such information about themselves, and that what to do with this information is up to them. In this "patient-centered" healthcare vision, the liberty of individuals and their self-determination are therefore central values. Individuals will not be dependent on doctors, and they will be able to take care of their health in an autonomous manner; ImSg is seen as empowering individuals in this respect. Interestingly, these "libertarian" values of freedom – based on the concept of personal property – are combined in researchers' scripts with "communitarian" values.[20] In fact, CIM researchers explicitly point to e-healthcare platforms in which users share their personal health information and data with other users in order to maximize the availability of interpretation tools.[21] This sort of open-source tool doesn't aim at accurate interpretation, but its value would be to provide a platform for healthy people to share information and knowledge concerning the relationship between data on immune system activity and the meaning of such data in terms of condition of health. The assumption is that people, while owning their personal information, need to share it with others in order to make sense of it. The moral connotation of this envisioned technology is therefore

---

[19] See Chap. 5 for a discussion on the moral aspects of the technical codes
[20] For a short introduction to libertarian and communitarian values see Swift (2001).
[21] See for example: www.patientslikeme.com

an interesting combination of libertarian ideas of freedom and autonomy together with a communitarian ideal of "sharing". Let's compare the moral meaning of this vision with the others.

The vision of ImSg as a tool for the diagnosis of disease enables a system in which doctors play a central role. Here ImSg is considered an empowering tool for the doctor, rather than the individual user. As explained by one respondent, a test is useful for doctors when it provides them with relevant information to make a clinical decision or to inform a patient (see Sect. 6.4.2). ImSg is expected to provide a clear-cut outcome that answers the questions of the doctor. In this context, ImSg is a good test not if it promotes individual autonomy or freedom, but if it promotes accuracy and salience within clinical practice.

Between these two visions there is a third vision, of ImSg in a "medical home" context. Here the same platform of DIY ImSg is embedded in a medical context: the technology is envisioned as a platform for monitoring the health of subjects from high-risk categories, such as old people, pregnant women or patients with chronic illnesses. In this scenario, the key users are not healthy people, but hospitals or primary care services. In fact, information provided by the system would help reduce costs for the care provider/payer (as explained by the director of the Innovation department, see Sect. 6.4.2). Controlling the health of those in high-risk groups is expected to result in measurable economic benefit. In this vision, ImSg is good not because it provides an accurate diagnostic test, but because it contributes to the profit of care providers and care payers (insurance companies). In this case, it is not the specificity of the test that is desirable, but the fact that it allows comprehensive, cheap monitoring. Furthermore, this vision aims at a "remote, but tighter" relationship between patients/citizens and care structures. The values of autonomy and personal freedom promoted in the first vision have here been replaced by the value of "controlling" patients in their homes.

The moral meaning thus substantially changes according to different contexts of use of ImSg. The differences in the moral connotation, however, are lost in general expectations of ImSg. By referring to the economic value of a health monitoring system for asymptomatic people, other moral connotations that are inscribed in co-existing technical codes remain unarticulated. Articulation of these diverse moral connotations shows that it would be a mistake to assess the desirability of ImSg as such; instead, the multiple artifacts, applications and values should be distinguished before the question of the desirability of ImSg can be addressed.

6.5.2 *Stakeholders and Normative Divergence*

Not only different technical codes have different moral connotations. Also, stakeholders and social actors hold diverse interests, values and preferences, which do not necessarily align with those of the ImSg developers. One example is that FDA regulation requires statistics on the specificity and sensitivity of each protein used in a biomarker test, in order to ensure that the test is accurate. According to this view, placing an entire protein on an array, rather than a randomly assembled peptide, is more accurate, because the protein is expected to reproduce the exact target of the immune system. However, as CIM researchers note, these protein arrays are both more expensive and more time consuming. Peptide arrays might be less accurate, but they are cheaper and faster to assemble, and can therefore be distributed more widely and – if they are to be used by healthy people – will be more cost-effective. This controversy is therefore not about the accuracy of the test, but is a question about what application is desirable. In this sense, ImSg is a good test for monitoring healthy people and a poor (inaccurate) test for use in specific diagnostics. However, the desirability of the test might also depend on the context. In some contexts, in which low cost is a significant added value (for instance, in a developing country), ImS may even be a desirable diagnostics test.

At a more general level, the question of whether "disruptive" or "incremental" innovations are better for society depends on what value systems are held by actors. On the one hand, some CIM researchers propose a new way of looking at problems in health care by addressing healthy populations. On the other, the NIH questions the relevance of CIM work for current medical practice, because it provides information that is too general and which doesn't address clinical questions. These two ways of assessing impacts on health care are based on normative differences. In one case, scientists should provide tools to address clinical needs: the clinical need comes first, inspiring research that addresses a specific question relevant to medical practice. Scientists – as one respondent with a clinical background explained – should look at the "differences that matter" for doctors in a clinical context. Here, "one size fits all" solutions are not desirable, and general patterns measured through whole populations might be misleading. In the other case, such patterns are valuable because they allow asymptomatic people to become aware of the status of their health and thus might guide them to change their lifestyle, or to take measures to improve their health condition. Ultimately, different actors hold divergent visions of what is desirable for society.

Finally, the three visions of ImSg described in the previous section (ImSg as DiY comprehensive monitoring, ImSg as diagnostic tool, ImSg in the medical home) raise different ethical and legal questions. In the case of the vision of ImSg as an online system of shared information, privacy issues may arise. In fact, in order to build a robust statistical correlation between antibody binding patterns on the array and a specific health condition, an extensive database has to be created by the ImSg developers. This database should include not only a sample and its analysis, but also environmental, geographical and personal data. How should this information be protected in the context of scientific research on ImSg?

In the case of the vision of ImSg as a tool for doctors, controversies could emerge around the liability of the physician in the case of a mistaken diagnosis. New genetic tests change the responsibility of the physician and the scope of her expertise. The doctor is now responsible for knowing and using the best available diagnostic tools for investigating her patients' symptoms and susceptibilities. However, genetic tests, marker interpretations, and risk probabilities are currently a very limited part of the training of medical doctors. What kind of skills would ImSg demand from a medical doctor? In the absence of any symptom in the patient, what kind of liability for offering a pre-symptomatic test can be imposed on a physician? Would a person who goes on to develop a chronic disease be successful in suing her GP for not having prescribed ImSg monitoring before the symptoms occurred? Such questions highlight not only that different technological platforms promote different systems of values, but that different stakeholders and social actors hold different systems of values. The plausibility of expectations of the desirability of ImSg should be assessed with respect to this diversity. Expectations of the desirability of ImSg tend to unify social actors under an abstract category of "American society". However, when this abstract category is opened up, different values and ideas of what is desirable emerge.

6.5.3 The Interactions Between Immunosignatures and Morality

The promise of ImSg is to solve the problem of high health care costs, empower individuals and doctors, and create a world without patients. However, ImSg will be more than a tool to solve social problems. Depending on the way they are designed, the interpretations provided around them, and their eventual context of use, Immunosignatures can be expected to do different things than those articulated in their "function".

In the vision of ImSg as a comprehensive monitoring system, scientists expect that the goal of ImSg is to reduce costs in health care through "educating" people. Sometimes the value of ImSg for educating individuals is articulated as a goal in and of itself because it contributes to "empowering" people. By displaying information about an individual's health parameters and showing how these parameters are linked to health, ImSg is expected by scientists and other experts to enable individuals to "see" these correlations and thus to become aware of them.[22] Thus ImSg appears to foster individual responsibility by providing users with relevant information about their body and general state of health: this information is then expected to "empower" people to take charge of their health. ImSg is considered a catalyst for changing people's mindsets about individual responsibility for health care. Just as people have changed their understanding of what is "private" in response to online social networks, they are similarly expected to adapt their daily habits to the use of a health monitoring system.

This line of reasoning builds on several assumptions. The first is that more information on their immune system makes people more knowledgeable. The second is that this knowledge empowers people to take action. The third assumption is that taking action allows you to take charge of (or responsibility for) your health, and that this enables people to be more in control of their lives (empowered). However,

---

[22] "Each individual could see their own biosignatures and the implications of these signatures on a regular basis. This should empower them to take more responsibility for their own health and stimulate their own scientific education on what this means" (From Selin 2007).

if we take into account that ImSg can be expected to "mediate"[23] the user's understanding of and action in the world, we can point out some rather richer dynamics of the interaction between morality and technology than found in the instrumental model in which ImSg is a means to empower people. In this context, the plausibility of these developments becomes less straightforward.

First, the information provided by ImSg does not simply increase knowledge regarding an individual's health condition. How the information retrieved through the use of the peptide array (see Appendix, Fig. 6.2) will be interpreted is still uncertain. This uncertainty also emerges in discussion of the best way for this information to be communicated to potential customers or clients. How will an average person "make sense" of this information? The problem, according to scientists, is that the concept of "immunosignatures" refers to a type of information that is fundamentally different from similar, currently available types of personalized information (for example genetic profiles). An "immunosignature" identifies an individual's health status at a particular point in time. In this respect, "immunosignatures" differ from "genetic signatures" (or genetic profiling): "genetic signatures" promise stable, constant information about an individual's genetic make-up, while immunosignatures vary in time and are "dynamic" in the sense that they reveal a snapshot of information about a continuously changing condition (the individual's immune response). In some ways the concept of *immunosignature* is closer to the concept of "signature" in our everyday usage. Like a handwritten signature, authenticity is not granted by absolute permanence and identity but by a *sameness* that is preserved over time. That sameness guarantees the link between a changing *token* and an immutable *type* or individual subject behind that. In this sense, the ontology introduced by the concept of *immunosignature* differs from the ontological assumptions of the concept of "genetic signature". Genetic information offers information about an individual's genetic make-up that can lead to a deterministic acceptance of character and behavioral traits. What does the personalized and yet *changing* immunosignature say about the *permanent* individual? ImSg introduces new concepts and perspectives on the relationship between bodily information and condition of health. The way in which ImSg "mediates" our way of experiencing health is therefore much richer than that suggested by an instrumental relationship (i.e. that ImSg information entails knowledge on health conditions).

These questions bring me to the analysis of the second assumption, according to which knowledge about immune system activity will enable people to take action. ImSg detect abnormal activity in individuals' immune system, indicating that one is asymptomatically sick. Experts envision this knowledge as empowering people to become proactive and therefore engage in lifestyle changes. However, it is not clear to what extent a person has power to take individual action. For example, clinical experts might be needed in order to prescribe a treatment. Furthermore, ImSg, as a comprehensive monitoring system, can be part of a toolbox for improving the efficiency of care providers. Here, then, the third assumption is questioned. In fact, when used in the context of the "medical home" ImSg is a tool for hospitals rather

---

[23] See Chap. 5.

than a tool for individuals. Therefore it *doesn't empower* individuals to be more in control of themselves, but is *a tool to control* individuals. In this sense, it is possible to question whether ImSg prescribe some form of social responsibility rather than enhancing individual empowerment.

Questioning the plausibility of linear expectations of how a technology will be a means to a desirable goal (or a solution to a problem) is important in order to avoid speculation. ImSg cannot be simply understood as a tool for achieving desirable goals: it carries specific ideas and concepts of health, disease, information and personalized diagnostics. These concepts will interact with existing concepts such that ImSg can be expected to do different things. Looking at the relations between this emerging technology and our current concepts and values is important when assessing expectations of Immunosignatures and their desirability.

6.6 Discussion

The reconstruction of the origin and history of the ImSg science and technology has pointed out the novelty of the concept of "immunosignatures". By using antibodies as a source of information about condition of health, and randomly assembled peptides to catch them, ImSg challenges two "paradigms" in the microarray and immunology world. For this reason, according to researchers, it triggers the disbelief of their peers. Despite the rhetoric around personalization, exemplified in the metaphor of Immunosignature as individually "signed" information, a closer look at laboratory practices shows that research focus is on the actual search for standardized "normal signatures". My analysis of expectations on the laboratory floors has also disclosed the different ways that ImSg is done in research practice, and suggested that alternative technologies, with different expected uses, co-exist behind the idea of ImSg. When situated in the actual scientific practice of "doing immunosignatures", the promise of ImSg acquires new dimensions. In particular, new scenarios emerge such as, the use of Immunosignatures as diagnostic tools for specific diseases. Together with ImSg researchers, I have articulated scenarios describing the potential contexts of the use of such technologies. On closer analysis, I pointed out that these scenarios reveal the promotion of different values to the ones mobilised in broad promises and expectations on ImSg (as the example of the video suggests).

This analysis aimed to show the applicability of the three-step plausibility framework. Instead of providing a "one-size-fits-all" method, the framework is tailored to each specific case study. As discussed in Sect. 6.2, I made different research design choices in addressing the Nanopil and Immunosignatures cases. These choices were dependent on a number of differences between the two cases. For instance, the Nanopil is an artifact developed by engineers under the guidance of an oncologist. The context of application is clearly specified (screening for colorectal cancer) and clinical practice is taken into account in the design of its different components. In contrast, Immunosignatures are "fingerprints" of the immune system retrieved through a specific array or microchip. The clinical application of this innovation is not yet fully defined, and different directions are being explored. In the first case, engineers are aiming to test and assemble different functional components; in the second, molecular biologists, biochemists and statisticians analyze biological samples and look for meaningful correlations. The different "epistemic cultures" (Knorr-Cetina 1999) of the two research groups require different methods for the analysis of expectations of the technology. If in the case of the Nanopil interactions with scientists were relatively limited – being based on a reconstruction of the different components and "pieces" of the artifact – for the Immunosignatures a longer and more intensive laboratory engagement was required in order to get acquainted with the science and technology. This longer laboratory engagement also allowed me to explore in more detail different possibilities for clinical applications, which were less well articulated than in the case of the Nanopil. In general, longer experience within the laboratory (2 or 3 months) better served the aim of analyzing expectations of the artifact. These differences between the two cases are also reflected within the findings. In fact, while in the case of the Nanopil discrepancies between the engineers' fictive script and current clinical practice were emphasized, in the case of Immunosignatures discrepancies within research practice were the center of attention.

It is important to stress that both cases presented in this book concern emerging technologies for molecular diagnostics. Can the plausibility framework that I am proposing here be applied to different types of emerging technologies? This framework is certainly not only meant for diagnostics or medical technologies, but could be applied to other technologies. The comparison between the Nanopil and Immunosignatures has shown that the specificity of the technology, its stage of development, and the context of innovation have to be taken into account. Despite this context dependence, three features are important for any case:

1. Two levels of analysis of expectations (of the technology and of its use) are conducted as a basis for an assessment of the plausibility of visions of the desirability of an emerging technology;
2. These two levels of analysis are approached with the same main strategies of "thickening" and "situating": expectations are explored within a situated (research, clinical or other social) practice, with the aim of enriching them with details that did not appear in widespread promises;
3. These analytic strategies aim to disclose the heterogeneity that is hidden within general visions. This assessment teases out different aspects of the scientific and technological platform, its usage, and the values rendered invisible in broader scenarios communicated to lay audiences. This unraveling fulfills the function of assessing the plausibility of these expectations, rather than evaluating their desirability.

Such plausibility assessments can be considered to have an "epistemological" rather than a "prescriptive" role because they aim to improve the conditions for developing knowledge about emerging technologies rather than offering a judgment on whether they are good or bad. This epistemological assessment is itself not

6.6 Discussion

value-free. As explained in Chap. 1, my approach holds a normative agenda of integrating reflection on the moral acceptability of emerging technologies into public deliberation on such technologies. According to this normative agenda, such reflection should be anchored in the current state of the art and in social practices and the wider moral landscape and yet be able to explore the ways in which emerging technologies might alter symbolic orders (see Chap. 2). Such an agenda is achieved by offering the plausibility assessment as an analytical tool for ethicists who want to engage in discussion on the moral acceptability of emerging technologies. Furthermore, the plausibility assessment can also help to build tools to facilitate stakeholder interactions within ad hoc spaces for deliberation, for instance in the context of technology assessment activities. The next chapter will elaborate on this aspect.

Appendix

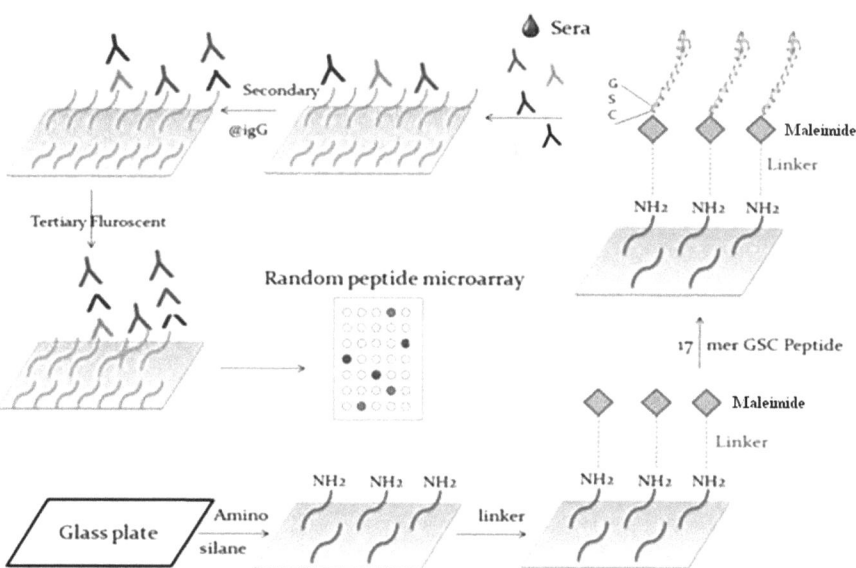

Fig. 6.1 The Immunosignaturing process: read from bottom *left* (Courtesy of Muskan Kukreja)

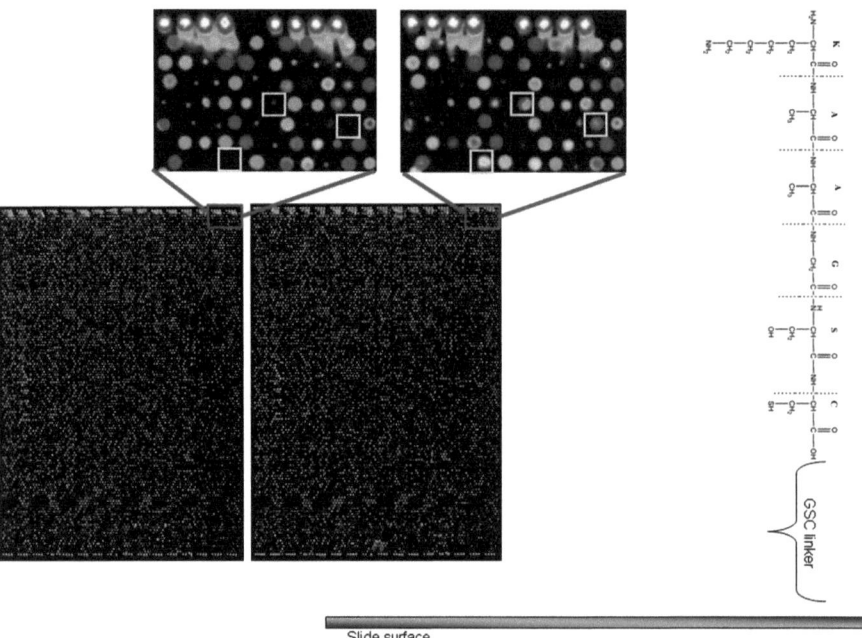

Fig. 6.2 From (Stafford et al. 2012) scan of the random peptide microarray (each colored spot correspond to a peptide previously printed on the array. The color depends on the intensity of the binding between each peptide and the antibodies in the sample – that has been run on the array)

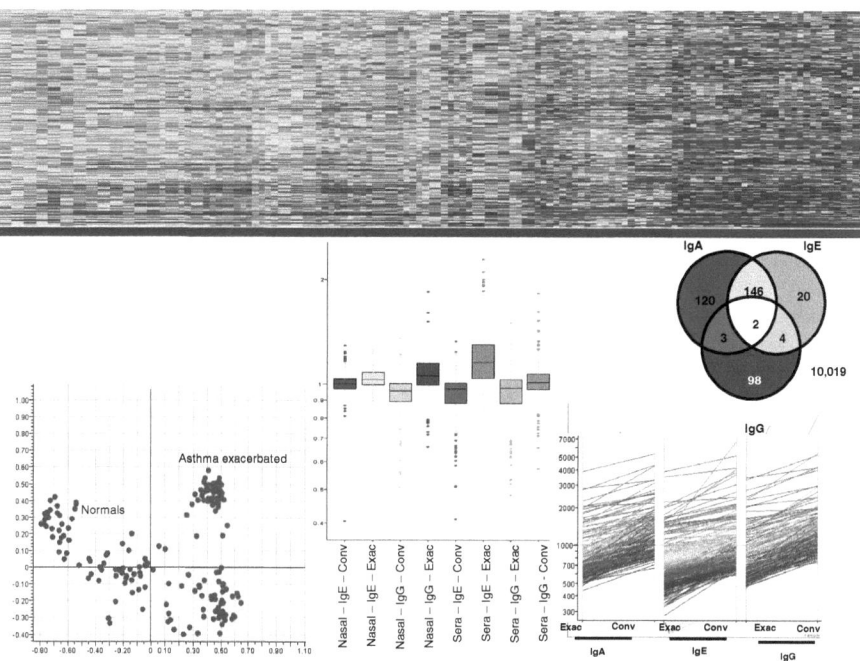

Fig. 6.3 (From Stafford et al. 2012) Five analytical representations of the data retrieved by the quantification of data showed in Fig. 4. After the measurement of the intensity with which the peptides lighten up, these data are analyzed with another software (also used for DNA microarray). Such software produces heatmaps, PCA and other graphs (see Fig. 6.3) which highlight different correlations and visualize patterns in the data. The *blue* color indicates a low binding of peptides and antibodies and the *red* indicates an high binding

References

Feenberg, A. 1995. *Alternative modernity: The technical turn in philosophy and social theory*. Berkeley: University of California Press.
Fisher, Erik. 2007. Ethnographic invention: Probing the capacity of laboratory decisions. *NanoEthics* 1(2): 155–165.
Fisher, E., and R. Mahajan. 2010. Embedding the humanities in engineering: Art, dialogue, and a laboratory. In *Trading zones and interactional expertise: Creating new kings of collaboration*, ed. Michael E. Gorman. Cambridge MA/London: The MIT Press.
Grin, J. 2000. Technology assessment as a tool for political judgement. In *Vision assessment: Shaping technology in 21st century society V*, ed. J. Grin and A. Grunwald, 9–33. Berlin: Springer.
Knorr-Cetina, K. 1999. *Epistemic cultures: How the sciences make knowledge*. Cambridge, MA: Harvard University Press.
Little, S. 2006. FDA regulations and novel molecular diagnostic tests. *Clinical Laboratory International* 7: 48–49.
Schuurbiers, Daan, and Erik Fisher. 2009. Lab-scale intervention. Science & society series on convergence research. *EMBO Reports* 10(5): 424–427.

Selin, Cynthia. 2007. *The future of medical diagnostics*. Scenario Development Workshop Workbook. Center for Nanotechnology in Society, Arizona State University, Tempe.

Selin, Cynthia. 2008. *The future of medical diagnostics*. Scenario Development Workshop Report, CNS-ASU Report #R08-0001. Center for Nanotechnology in Society, Arizona State University, Tempe.

Stafford, P., R. Halperin, J.B. Legutki, D.M. Magee, J. Galgiani, and S.A. Johnston. 2012. Physical characterization of the "Immunosignaturing Effect". *Molecular & Cellular Proteomics* 11(4).

Star, S.L., and J.R. Griesemer. 1989. Institutional ecology, 'translations' and boundary objects: Amateurs and professionals in Berkeley's Museum of Vertebrate Zoology, 1907–39. *Social Studies of Science* 19(3): 387–420.

Swift, A. 2001. *Political philosophy: A beginners' guide for students and statesmen*. Oxford: Polity.

Tedlock, Barbara. 2000. Ethnography and ethnographic representation. In *Handbook of qualitative research*, ed. Norman K. Denzin and Yvonna S. Lincoln, 455–486. Los Angeles/London/New Delhi/Singapore/Washington DC: Sage.

van der Burg, Simone, and Tsjalling Swierstra. 2013. *Ethics on the laboratory floor*. Basingstoke: Palgrave Macmillan.

Chapter 7
Scenarios as "Grounded Explorations". Designing Tools for Discussing the Desirability of Emerging Technologies

> *[We] really are only able to, and need to, question our basic assumptions in the moment when we collide with an element of the complexity of our life, a tear in the routine of experience that requires us to rethink things in order that we might progress along our current (or any other alternate) course.*
> (McGee 2002: 112)

Abstract This chapter addresses the question of how the analysis of expectations' plausibility, described in the previous chapters, contributes to the goal of fostering a democratic deliberation on the normative acceptability of emerging technologies (outlined in Chap. 1). Building on pragmatist ethics approaches, the normative ideal of democratic deliberations around science and technology is outlined as an attempt to include different perspectives in the deliberative process; articulate the reasons, meanings and assumptions behind a problem; and explore possible scenarios of how new technologies change our moral concepts and vocabularies. In this context, "moral imagination" is presented as a way of broadening democratic deliberation exercises to create spaces for discussions of "ideas of good" and moral values. After reviewing some types of scenarios that have been developed as tools for discussing the social and ethical issues of emerging technologies (namely, "socio-technical" scenarios and "techno-ethical" scenarios and vignettes), indicating some of their limitations, the chapter addresses the issue of using scenarios to trigger the moral imagination of technology developers. It does so by discussing two exercises based on the analysis of expectations' plausibility conducted in the previous chapters of this book that have been designed in order to produce a "grounded" and "exploratory" discussion with the developers of the Nanopil and Immunosignatures.

Keywords Pragmatist ethics • Democratic deliberation • Moral imagination • Techno-ethical scenarios • Techno-moral vignettes • Workshops with scientists and technology developers

7.1 In Search of a Normative Framework

Expectations and visions of emerging technologies have a normative content that needs to be articulated. Based on the literature and conceptual framework discussed in Chaps. 2, 3, 4 and 5 described a methodological approach to articulate visions' normative content. This approach was inspired by the concept of plausibility and guided by the idea that normative contents and moral values can be articulated when exploring material practices of technological design, existing societal practices, stakeholders' worldviews and trends of co-evolution of technology and morality. How does this analysis of expectations' plausibility contribute to the goal, discussed in Chap. 1, of fostering a democratic deliberation on the normative acceptability of emerging technologies which guarantees a plurality of views? This chapter addresses this question by offering two examples of "deliberative exercises" with technology developers that have been designed in line with a pragmatist ethics approach. Why pragmatist ethics? As explained in the following, pragmatist ethics offers a normative framework based on democratic deliberation that seems to resonate with TA's implicit normativity.

The enterprise of assessing emerging technologies demands a normative assessment, that is, an evaluation of the technology at stake. As discussed in Chaps. 1 and 2, Western societies have increasingly acknowledged the need for democratic procedures that would include different views and perspectives in policy deliberations around science and technology. Such a broadening of participation has also come with a broadening of relevant issues for TA. While economic values and health and safety were of primary concern when TA offices were initially established, it later became clear that broader social values should also be considered in such evaluations. As we saw in Chap. 1, some second-generation TA approaches seem to align with "deliberative democracy" ideals by aiming to create the conditions for participants to produce reasonable and well-informed decisions in which they take into account the positions of the other participants (Sect. 1.2). For example, Constructive Technology Assessment (CTA) aims to improve the conditions for technology development at an early stage of the innovation process and this is done by providing stakeholders with tools to incorporate into decision making processes knowledge about socio-technical dynamics (Schot and Rip 1997). TA practices have, however been criticized for not doing justice to their normative ideals and not engaging in assessments that would explore the normative positions at stake in deliberations on emerging technologies to articulate implicit moral issues, or to foster discussion of the interplay of new technologies with moral vocabularies. Some TA scholars and practitioners have highlighted the need for TA activities to focus on moral values as well, pleading for an integration of ethical inquiry into TA that would address such "normative deficit". In these contexts, however, traditional normative ethics, which provide prescriptive judgments on emerging technologies from pre-conceived normative theories, fall short as they cannot account for a plurality of stakeholders' worldviews and normative positions. Applied ethics' normative appraisals of new technologies are often based on traditional ethical theories (deontological, consequentialist or virtue-ethics based) and the moral principles on which they are grounded. The outcome of such an appraisal is in the "shall/shall not" form. The programmatic work, by Keulartz and colleagues (2002, see also Chap. 1) suggests that a pragmatist ethics approach provides a valuable normative and conceptual framework that can address this gap between normative ethical theories and TA which is articulated around the ideal of democratic deliberation.

7.1.1 Democratic Deliberation as a Normative Ideal: A Pragmatist Approach

According to pragmatist ethics,[1] values are not considered to be universal and transcendent rather, they are understood as situated in cultures, practices, times, and spaces (James 1956[1897]; Dewey [1929]; Rorty [1989]). In this sense, pragmatists share a sense for context in so far that they argue that there are no prior universally valid normative principles on which we can ground and justify our actions. On the contrary, some values are "right" because they work in a particular context.[2] According to John Dewey's pragmatist philosophy, norms and values are solutions to practical problems [1929]. In particular, morality is the set of values, norms and standards of good behavior that are shared in a certain society. Ethics is "reflective morality", that is, the explicit discussion and questioning of moral principles and models (Dewey [1992]). Ethical reflection happens as a discursive practice that assesses what moral actions to take: ethics is "a reflective discipline for factual morals" that has a practical relevance *"only if technology decisions involve moral conflicts"* (Grunwald 2004: 180 based on Habermas 1973). In line with a pragmatist epistemology, norms and values are solutions to practical problems. Dewey's contextualism emphasizes that morality is the uncontroversial ground of our behavior and is questioned only in situated, real, *problematic situations*. When a concrete problem emerges, the guiding principles of our actions need to be discussed and eventually adapted to the new situation. This means that "morality" is not always consistent.

Characterized by a contextual and dynamic approach to morality and ethics, pragmatist ethics does not offer a grand substantive moral theory that can be used to evaluate emerging technologies, but rather focuses on "problematic situations". This does not mean that "anything goes". Moral pluralism is something different from moral relativism (Appiah 2010). The concept of "right" is set as a regulative ideal in the practice of democratic deliberation (Habermas 1990). The process of finding workable solutions for problematic situations is a process of collective learning

---

[1] Although pragmatist philosophers are hard to bring under a common denominator, some similarities are evident. For an overview of the classic pragmatist philosophers see Ayer 1968; Rorty 1982; Murphy 1990; Putnam 1987.

[2] This has been referred to as the anti-foundationalist thesis that is shared in pragmatist ethics as well as in pragmatist epistemology (see Peirce 1960a[1877], b[1905])

aimed at achieving a regulative ideal of "creative democracy". Keulartz and colleagues highlight that "a pragmatist ethics is more process – than product – oriented" (2002: 15). Rather than trying to offer solutions, this approach aims at "facilitating public debate and political decision-making on emergent moral problems" (ibid.). Pragmatist ethics contributes to the process of deliberating on emerging technologies by facilitating public and political debate and by ensuring that all relevant groups contribute to the discussion by having an opportunity to articulate their views.

The underlying theory is based on the Habermasian idea that a fair and non-coercive moral discourse creates the best conditions for participants to account for their positions, defend them, respond to critiques and learn from the positions of others (Habermas 1990). However, some authors reflecting on the potential of pragmatist ethics for a technological culture also point out the limitations of this Habermasian perspective. In fact, Habermas' ideal space of inquiry is an abstraction that is, in practice, actualized in interactions among subjects dominated by power-relations and ideologies. Building on a tradition inspired by Michel Foucault, some authors emphasize the importance of making the meanings explicit, and showing "to what people who use and apply concepts are committed" (De Vries 2002: 154). This stresses both the importance of collecting different perspectives as well as the need for critical analysis of foundational presuppositions. Rather than aiming at making the right decision, this approach assumes that the process of articulating the problems and analyzing the epistemological conditions for deliberation is important and desirable as it challenges implicit moral assumptions. Such critical and creative analysis is crucial for deliberation. As Zwart elegantly explains:

> By studying the conditions that allow certain problems and solutions, as well as certain principles and concepts to emerge, we may become more aware of the factors that actually guide our thoughts and actions in the present, and this may prepare us for making our own concrete choices with regard to the present, although the choices are never determined by our philosophical principles, nor can they be deduced from them or warranted by them. (Zwart 2002: 39)

In this sense, such an approach should be considered as a "diagnostic practice", a screening of the problem, rather than as a treatment.

Improving the process of democratic deliberation requires not only the inclusion of different groups, but also requires a broadening of the issues discussed in the deliberative process. As Thompson (2002) remarks, in order to be truly consistent with the normative goal of deliberative democracy, pragmatist ethicists can maintain the balance between a "liberal" approach that focuses on the procedures for decision-making and aims at compromise or consensus among participants, and a "republican" approach that focuses on the formation and quality of opinions and interests. In the republican approach, consensus for decision-making is a secondary goal. The main goal lies in the capture of all relevant arguments and alternatives. Differences and confrontations are kept lively in order for the process of preference formation to become more robust. In this sense, pragmatist ethics offers a normative background that promotes deliberation of emerging technologies in which different perspectives and groups play an active role and in which a broad range of issues and topics are explored.

7.1 In Search of a Normative Framework

Pragmatist ethics also promotes a more substantial intervention in the process of moral deliberation on emerging technologies. Such an intervention can be done in three ways (Keulartz et al. 2002). First, by defining the nature of the problem. In doing this, the ethicist pays attention to how a solution in one problem area can be transposed and translated in another area (as we saw in the case of the birth control pill). In doing this, it is important to look at the two sides of the technology-morality relationship: on the one side, technology shifts the perception of moral problems, and on the other side moral ideas influence the perception of technological problems. Moral deliberation can be enhanced, second, by sketching possible future scenarios. Such scenarios provide a narrative context for identifying emerging "problematic situations" at an early stage of technological development. Third, they help in anticipating how problems addressed by an emerging technology may create other "problematic situations" in different contexts. In doing this, the ethicist should look at problems beyond the currently accepted values promoted by the technology and propose scenarios that address questions about possible future worlds that are disclosed by certain technologies, taking account of how they change our view of nature, others and ourselves, shifts in patterns and responsibilities and in relations of power (*ibid*, 260).

These scenarios make actors more reflexive about their normative vocabularies and question whether these morally accepted ways of speaking require revision. This last point is linked to the third substantive intervention of the pragmatist ethicist in the deliberative process on emerging technologies: the ethicist can engage in the development of new moral vocabularies thus avoiding dichotomous and polarized positions in ethical debates. This is achieved by articulating nuances and degrees rather than proposing ontological categories to analyze moral conflict. At times, the ethicists may be required to redefine concepts by looking at the problem from different perspectives.

Pragmatist ethics provides a normative framework that encounters the sociological, contextual, process oriented goals of some forms of TA (specifically CTA[3]). Being normatively grounded in the ideal and practice of deliberative democracy, it offers tools that enable an exploration and discussion of values in technology and social contexts. By being sensitive to specific contexts and aiming at situated workable solutions rather than all-encompassing prescriptive recommendations, this approach is in line with the process-oriented, pluralist and democratic approach of CTA.[4] By focusing on the relations between technology and morality and the normative assumptions that are hidden within deliberative practices, pragmatist ethics offers a framework that aims to create the preconditions for a democratic discussion on the moral acceptability of emerging technologies.[5]

---

[3] Constructive Technology Assessment (CTA) is explained in Chap. 1.

[4] For an incorporation of a pragmatist ethics approach in CTA see Shelley Egan 2011, and Krabbenborg 2013.

[5] It should be pointed out that according to Brom and Est 2011 "discursive or argumentative TA" shares a similar normative orientation.

7.1.2 Triggering Moral Imagination

As explained above, authors who acknowledge the role of a pragmatist ethics approach for a technological culture argue that Habermasian approaches, which focus primarily on the procedural aspects of a democratic deliberation, are insufficient. According to them, such interventions should also be "substantive" if they are to facilitate public debate and political decision-making. Such interventions help open up perspectives and explore moral conflicts. The notion of "moral imagination", inspired by John Dewey, plays an important role in this more substantial type of intervention. Moral imagination is a cognitive human competence important for decision-making (Fesmire 2003; Coeckelbergh 2007); specifically, it is significant when we make decisions according to particular epistemic and moral criteria. We decide (or not) to do something on the basis of what we (think we) want: on what we know about the world: or on what we think is right to do. In this sense, deliberation is a process of moral reasoning. Making the moral content of this reasoning explicit should, therefore, improve the process. Moral imagination plays an important role in this process, because it allows people to project themselves and their actions into the future, to "feel" how it would be. This capacity for "putting yourself in someone's else shoes" – imagining how you would feel and act in a potential future situation – is a key reflexive skill for moral reasoning.

Moral imagination is therefore an important skill to develop in those who make decisions around emerging technologies (Burg 2009, 2010). Technology developers, policy makers or investors need to make daily decisions that concern future technologies that determine their nature and applications. Decisions are always made in specific contexts: in the following, I will focus on tools to broaden moral imagination in the deliberative context of the "laboratory". When technology developers explore different possibilities for applying a scientific and technological innovation in a clinical context, they have to discuss these possibilities and make a decision on what is the best direction. In such a decision making process, different values and considerations play a role – for example questions of feasibility or profit. The expectation that Immunosignatures[6] will change people's standards of privacy guides has, for example, guided (or you could say 'informed) scientists' visions of how it will become embedded into society. Here the ability to imagine how you would feel and act in the future is an important skill.[7] After reviewing some of the types of scenarios that have been developed as tools for discussion of the social and ethical implications of emerging technologies and their limitations (namely, "socio-technical" scenarios and "techno-ethical" scenarios and vignettes), indicating some of their limitations (Sect. 7.3), the chapter considers how scenarios are, or can be, used to trigger the moral imagination of technology developers. It does so by

---

[6] The case of Immunosignatures is discussed in Chap. 6.

[7] It has been argued that since broader social and normative considerations play an important role in decisions made on the laboratory floor, then making scientists reflexive about the role of this type of consideration in their decision making process is important in order to modulate technology development in a desirable way (Fisher 2007).

7.2 Scenarios as Tools to Foster Moral Imagination

discussing two exercises that, based on the analysis of expectations' plausibility, conducted in the previous chapters of this book, have been designed to engage in "grounded" and "exploratory" discussion with the developers of the Immunosignatures (Sect. 7.4.1) and the Nanopil (Sect. 7.4.2).

7.2 Scenarios as Tools to Foster Moral Imagination

Future "scenarios" constitute suitable tools for training and triggering the moral imagination so as to help deal with decisions on emerging technologies. Scenarios are "creative redescriptions" (Keulartz et al. 2002) of "various competing possible lines of action and courses of conduct". They offer a plurality of possible futures and, in this way, present a number of alternative ways to address problems. Within the pragmatist perspective, scenarios are particularly appropriate to use when deliberating emerging technologies because we have to deal with "new problems, for which existing rules and routines are not adequate" (ibid.: 259). Scenarios are useful tools for stakeholder reflection on the desirability of emerging technologies because they enable them to try out different plans in their minds, and to imagine the consequences of their choices and actions. More specifically, scenarios enable stakeholders to imagine how concepts, systems of values, and responsibilities (in a word, our morality) can evolve together with new technologies. In this way, the desirability of emerging technologies is not discussed using old concepts and vocabularies, but in a manner that enables their novel and creative role to be explored.

> Which possible future worlds are disclosed by certain technologies? How do they change our view of nature, each other and ourselves? What shifts in patterns of responsibilities and in relations of power do they bring about? Next, it can then be investigated whether our normative vocabularies – our usual, morally accepted and socially sanctioned ways of speaking- are still adequate in light of these changes and shifts, or whether they require revision. (ibid.: 260)

Scenario narratives require stakeholders to step out of their current routine and engage with moral dilemmas and problems. In these hypothetical "problematic situations", stakeholders are asked to reflect on their moral vocabularies and to adapt them to new problems. Pragmatist ethics acknowledges the importance of scenarios, but doesn't provide a toolkit for writing and using them. These aspects have been investigated in other contexts.

Foresight (or Future) Studies have developed scenario methodologies to explore possible futures, support the policy process and engage stakeholders in deliberative exercises (Brown et al. 2001; Schomberg et al. 2005). Several types of scenarios have been developed in order to help trigger discussion amongst stakeholders in the context of Technology Assessment. Constructive Technology Assessment (CTA)[8] has developed several tools in this direction. CTA workshops are interactive, heterogeneous settings that create a "protected space" for stakeholders to reflect on

---
[8] For an extensive explanation of Constructive Technology Assessment see Chap. 1.

technological development, position themselves with respect to others, probe each other's worlds, articulate their perspectives, and listen to other points of view. Stakeholders therefore have to be able to articulate their position, defend their arguments, and criticize and learn from others.

Socio-technical scenarios can be used as inputs into these discussions. They are fictional narratives that, starting from an existing socio-technical configuration, describe potential future developments and dynamics triggered by decisions by different stakeholders (Rip and te Kulve 2008). Building on previous analytic research on socio-technical dynamics, these scenarios describe how technological innovation breaks up the existing socio-technical order and requires actors to "re-align" in different combinations (Robinson 2010). Presented to stakeholders during a CTA workshops, scenarios are used as tools to broaden stakeholder understanding of socio-technical dynamics and to develop reflexivity on one's own role in shaping future configurations, as well as of stakeholder interdependence. Such an understanding is expected to broaden actors' perspectives and enable them to play an active and more aware role in the innovation process. In this sense, they offer both a window of opportunity for modulating decision-making and a strategy for moving in a desired direction (Robinson 2010): "Broadening perspectives and providing insights in sociotechnical dynamics enables actors to do better in their normal working environment and can eventually contribute towards more desirable paths" (van Merkerk and Smits 2008: 329). Although a discussion of the desirability of an emerging technology can be triggered by this interactive exercise, this is not its primary aim. As has been noted: "Scenarios offered entrance points for ethical discussion, but these were rarely taken up in the workshops; other issues such as start-up firms and regulation dominated" (Shelley Egan 2011, reporting Robinson 2010).

With the specific goal to foster ethical reflexivity among scientists, other tools, namely *argumentative scenarios,* have been developed and implemented (Shelley-Egan 2011). Based on Stephen Toulmin's Model of Argumentation and on interviews with scientists, these scenarios show how a *claim* is justified by a *warrant* (based on an ethical theory), and is counter-argued with a *reservation*. These tools have been introduced into focus groups with scientists in order to develop their understanding of the argumentative structure of moral dilemmas around scientists' responsibility and accountability. These tools share with socio-technical scenarios the goal of stimulating scientists to discuss their positions, specifically by focusing on ethical arguments. As schematic representations of the functions of and relations between claims, warrants and reservations, these tools can enhance actor reflexivity about the discursive structure of ethical arguments. However, they do not trigger actors' moral imaginations.

This goal is met in the narrative form of *techno-ethical scenarios* (Swierstra et al. 2009a, b; Stemerding et al. 2010). These are fictional narratives of future ethical controversies around new science and technology. These scenarios are built according to a three step approach in which (1) the current moral landscape is outlined; (2) the technology is introduced and some potential ethical controversies are played out; and (3) some closure to these controversies is given through (fictive, but plausible) technical, regulative or organizational solutions (Boenink et al. 2010). These

7.2 Scenarios as Tools to Foster Moral Imagination

three steps can be repeated several times so as to articulate further future controversies and closures. These scenarios provide a tool for the anticipation of ethical controversies around a new technology and for the exploration of the dynamics of interaction between current morality and new technologies. These tools have been used in CTA workshops to initiate a discussion on the roles and responsibilities of different stakeholders in the development of a specific technology (Lithium Chip and BAN technology) (Krabbenborg 2013).

Techno-moral vignettes[9] have been used as tools to initiate discussion of potential "soft impacts" of emerging technologies; that is, the impacts of the technology on forms of life, concepts, and morality. These impacts are considered important by the lay public and should therefore be included in democratic deliberation on emerging technologies. However, some social actors consider these impacts "soft" because they are not clearly quantifiable, objective risks (Swierstra and te Molder 2012). For this reason, such impacts are often neglected in public discussions. Like scenarios, vignettes are narratives that focus on potential impacts of new technologies on ways of life and which give accounts of techno-moral change. However, unlike scenarios they do not "offer a line of action and course of conduct". In fact, scenarios, like movies, are a narrative form that describes a temporal unraveling of events. In this way they draw attention to alternative pathways and dilemmas that can emerge as a consequence of particular decisions. In contrast vignettes are more similar to photographs: they are "snapshots" of the future (van Notten et al. 2003). Unlike scenarios, vignettes do not show causal connections between events or explanations of how a certain situation occurs. While scenarios take stakeholders by hand to show the potential co-evolution of technology and morality step-by-step, showing causal pathways between events, vignettes describe a future state of affairs.

Techno-ethical scenarios and techno-moral vignettes provide useful tools for pragmatist ethics' goal of improving the deliberative process by triggering the moral imagination. An important aspect of these scenarios is that they are "controlled speculations" (Rip and te Kulve 2008); that is, the scenarios do not claim to have predictive power, but are based on empirical study of the context and dynamics of the innovation process. The "controlled" or "grounded" character of scenarios makes them a good tool for triggering discussions in which participants are "experts" or stakeholders who have been asked to reflect on the development of a specific technology. However, scenarios have also a speculative aspect, because they "explore" a potential future in order to trigger discussion.[10] Assessing the plausibility

---

[9] These tools were initially developed within the project "Vignettes and scenarios", funded by the Nanopodium program (initiative of the Dutch Committee for the Societal Dialogue on nanotechnologies). See http://www.nanopodium.nl/CieMDN/projecten/overig/vignetten_en_scenarios. Vignettes were further elaborated within a project coordinated by the Rathenau Institute: Synbio Futures. This project aimed to engage youth political organizations in debate on synthetic biology. Vignettes were developed and employed as a tool to trigger such debate. For more information on techno-moral vignettes and their role in this project see http://www.rathenau.nl/themas/project/synthetische-biologie/synbio-futures.html

[10] The discussion of speculative ethics and the middle ground between "here and now" ethics and "exploratory ethics" was introduced in Chap. 1.

of expectations contributes to the aim of constructing techno-ethical scenarios/vignettes as "grounded explorations" of future problematic situations.

7.3 Plausible Scenarios for "Grounded Explorations"

The three steps of the plausibility assessment help in writing scenarios that are grounded in the current state of the (technical) art, social practices, and moral landscapes, and yet explore possible future interactions between technology and morality. First, analysis of expectations of the artifact articulates the current state of the art, including technical challenges and scientific uncertainties. Description of the artifact's technical components provides better understanding of the technology (for example, the laxative condition, in the case of the Nanopil), and excludes some scenarios (for example, DNA molecules' short lifespan excludes scenarios which envision the Nanopil being available at grocery stores). Furthermore, the articulation of competing technical alternatives allows the selection of a potential future "fork" in the decision making process (for example the wireless vs. blue bolus signal communication). Upstream discussion of the moral acceptability of these two alternatives might support a future deliberative process around them.

Second, the analysis of expectations of use articulates the "fictive script" of technology developers such that it can be compared to perspectives of other social actors. In this way, current practitioners point out conditions for the success of the emerging technology in a particular context, and highly speculative scenarios of widespread and unconditional social use are avoided. Description of current practice articulates existing social arrangements which might hinder some design choices and promote others (for example, some existing practices of the national screening program are not compatible with the blue bolus system).

These analyses guide the assessment of the plausibility of expectations that an emerging technology will bring about a desirable world. Such assessments explore possible value conflicts and divergences which can then be the starting point for storylines for techno-ethical scenarios/vignettes. Firstly they can point out the conflicts between values inscribed in different designs for the same artifact. This distinction is important in terms of not clustering different ideas of the artifact into one scenario. Second, differences between systems of values shared by technology developers and those of other social actors are articulated. In this way, scenarios can explore the ethical controversies that might arise around a certain technology. Third, such analysis examines the mediating role an emerging technology may take in a certain social practice. Scenarios can thereby point out the potential future interplay between technology and morality. In general, the plausibility assessment that I proposed excludes some scenarios as implausible while at the same time identifying conditions, conflicts and options that affect the desirability of the emerging technology.

Vignettes and scenarios are vehicles for feeding the results of the plausibility assessment back to stakeholders and thus triggering a discussion of the desirability of emerging technologies that is both grounded and exploratory. This balance between the goal of groundedness and exploration deserves some attention. Assessing the plausibility of expectations requires one to analyze the state of the art in technology and research practice, social practices, and the normative positions of different stakeholders. This exercise provides a solid ground for non-speculative scenarios and vignettes; however, this type of grounding is not enough. In fact, scenarios and vignettes are not *predictions* of possible futures, but tools to trigger the moral imaginations of discussants in ad hoc exercises (such as workshops) such that they encourage a discussion of the desirability of emerging technologies. In order to achieve this goal, scenarios should be considered plausible by their readers. Scenarios that are considered implausible by the discussants will be easily dismissed. However, the plausibility of these scenarios is not an objective criterion. When writing them, the epistemic background of the target readers or discussants should be taken into account. What do they know about the technology at stake, or about the social context? What will they consider plausible? Scenario plausibility has to be measured with the yardstick of the target audience. Furthermore, the scenarios/vignettes should be anchored to what is considered relevant and important by target readers. In this way, they will trigger the interest of the discussant. However, it is important to highlight that if the goal of the whole exercise is to broaden discussion and explore potential problematic situations, then vignettes and scenarios also have to be capable of stretching the borders of what is considered relevant. An ideal intervention should therefore broaden the topics and issues considered relevant by stakeholders to deliberation on emerging technologies.

The plausibility assessments of expectations of the Nanopil and Immunosignatures presented in Chaps. 3, 4, 5 and 6 respectively have guided the construction of techno-ethical scenarios and techno-moral vignettes based on these technologies. These scenarios/vignettes should be considered tools for the facilitation of stakeholder exploration of the desirability of emerging technologies while avoiding highly speculative discussion. I implemented two exercises to investigate the utility of these tools, which were developed in the context of the plausibility assessments. In this study, scenarios and vignettes based on the Immunosignatures and Nanopil were used to broaden discussion of ethical concerns with technology developers of these two technologies.

7.4 Techno-Moral Vignettes and Scenarios in Action

The main goal of the two exercises was to explore the possibilities of using plausible techno-ethical scenarios and techno-moral vignettes in the context of the protected space of an interactive stakeholder workshop. My research question was: how do these tools foster discussion of the desirability of an emerging technology?

In order to answer this question I organized two workshops with the developers of the Nanopil and Immunosignatures in which the potential ethical and social implications of these technologies were a central focus of discussion. The techno-ethical scenarios and vignettes were used as tools for broadening participants' ethical considerations. Since these exercises were conceived as a pilot, to help understand the feasibility of the use of plausible scenarios and vignettes in an interactive context, I opted for a homogenous group of participants. Another reason for this was the reluctance of the developers to engage in multi-stakeholder workshops given the early stage of technology development.

These two exercises had different set-ups. Because the context for Immunosignature use is not yet well defined, with technology developers swinging between two very different visions of its application, the scenarios seemed a good tool to explore these different paths. In the case of the Nanopil, context of use – as screening for colorectal cancer – was more clearly defined, and could be described and explored within the short narrative of a vignette.

The success of these tools in facilitating discussion of the desirability of emerging technologies depends on how plausible they appear to workshop participants and on their effects on discussion. In evaluating the exploratory role of the vignettes/scenarios I have therefore addressed a number of questions: do they succeed in creating suspension of disbelief amongst participants? Are participants willing to discuss the issues that are raised in the scenarios and vignettes? Did scenarios/vignettes trigger/broaden discussion of the desirability of the technology at stake? If so, in what ways? And what were their ultimate effects on participants?

7.4.1 Workshop 1: Immunosignatures

The first workshop engaged some of the scientists involved in developing Immunosignatures (ImSg from now on) at Arizona State University. One year after the beginning of my 3-month laboratory engagement at the Centre for Innovations in Medicine (see Chap. 6), I invited the researchers involved in the ImSg project for an afternoon workshop. Amongst the participants were the director of the center, three senior researchers (assistant/associate professors), four graduate students, and two technicians. A colleague, also involved in the composition of scenarios, participated in the workshop so as to share the role of probing participants. As preparation for the workshop participants received two scenarios exploring two possible paths in the development of ImSg. Both scenarios had two parts: a first phase exploring the short-term future, and a second phase covering a longer-term future. The workshop was organized in five rounds. During rounds 1 and 2, participants were invited to read and discuss phase 1 of the two scenarios. During rounds 3 and 4, phase 2 of both scenarios was distributed and discussed. The last round was devoted to the evaluation of the workshop and scenarios.

Techno-Ethical Scenarios The two scenarios (reported in the Appendix of this chapter) are explorations of potential pathways for the co-evolution of ImSg (and their social embedding) with values, obligations and responsibilities. In particular, they aim to explore the differences between the two visions around ImSg pointed out in the plausibility assessment of Chap. 6: a patient-centered vision, and a

7.4 Techno-Moral Vignettes and Scenarios in Action

doctor-centered vision. At the end of my laboratory engagement I had presented this double vision to the Center's researchers during a group meeting. The Center's director found this distinction helpful – even "inspiring" – because it pointed out a "fork" for a decision that they have to make. For this reason, I thought it relevant to explore this fork further in the scenario workshop. Rather than taking the visions as two phases in the development of ImSg (as researchers at the Center tended to do), the two scenarios explore them as different developments. Both scenarios begin with a similar description of the current situation, in which preventive and personalized trends in healthcare appear as desirable solutions to unsustainable healthcare costs. The two storylines then develop in different directions, showing how the two visions held at the Centre might trigger different dynamics of interaction between technology, society and morality. They also suggest ethical dilemmas and controversies which might emerge along each path.

Scenario 1 focuses on the "patient-centered model", in which pre-symptomatic diagnosis is expected to empower individuals. According to this vision, people do not need doctors and are able to control their health autonomously. By enabling individuals to become more aware of their health, ImSg is expected to make them more responsible towards it. The scenario shows how these promises can be integrated with the promises of existing online platforms for information sharing among patients. This scenario thus explores the possibility that the system is used to promote "community" and "solidarity" values rather than only "liberal" and "private" ones, as the promise of "empowering the individual" might suggest. This ambiguity is expressed in a controversy about privacy rights around personal information. Closure is provided by the decision, by the authorities, to make patients' personal information available online for scientific research purposes. Phase 2 of the scenario describes how ImSg, conceived as a way of empowering individuals through freedom from doctors and hospitals, might instead become a tool for exerting control over them. By its inclusion in healthcare programs, ImSg can be used by institutions to invite citizens to take up responsible, healthy behaviors, with the goal of reducing costs in healthcare. However, this also burdens individuals with new responsibilities and duties.

Scenario 2 develops the vision of a "doctor-mediated model", in which personalized diagnostics are expected to improve healthcare services. Here ImSg provides a tool for improving current diagnosis practices. Because it aims to improve current practice, researchers expect that this vision will be easier to realize in the short-term. The scenario shows that the promise of personalized diagnostics can have its own development, rather than merely being a phase of development of tools for pre-symptomatic diagnosis. The scenario shows ImSg – when used by physicians to diagnose specific health conditions, such as Valley Fever – changing medical practice. In this context, ethical questions arise around criteria for a "good" diagnostic examination. Conflict between traditional diagnostic techniques (based on discussion of symptoms and physical examination) and molecular diagnostics are voiced by primary care doctors, who are burdened with new responsibilities for interpreting results and for acquiring scientific knowledge, while also being increasingly time limited. Phase 2 of this scenario describes how ImSg for tracking cancer recurrence becomes included in hospital remote-care programs. These programs, driven by the economics of cost reduction, raise the question of what "personal care" is when clinical personnel disappear. While the autonomy of patients is fostered, and they are free to decide how to act with respect to their own health, they are left on their own without human contact with their physician.

Opportunities for Exploration The discussion started with comments on scenario 1. One participant reflected on the fact that if you have to pay every time you want to have your immunosignature tested, "this can hinder the development of the technology". ImSg should become open-source in order for people to become acquainted with it, as in the case of social networks such as Facebook. Questions of privacy and the possible misuse of data were addressed by another participant. Later on, the same participant explains;

> I wouldn't want to have my immunosignatures public […] it's a personal choice; maybe some people want and some other people don't.

The example of the social network Facebook was again brought up to point out how people's attitudes and behaviors around privacy issues change over time. Another participant explained that, in this way, "everyone can work on this information" and "people can co-operate" to interpret information. Besides keeping costs low, this strategy would enable a "massive database" that would help make immunosignature interpretation more reliable (because it will be grounded on a large statistical base).

Another participant explained that he had never thought about the issue of privacy: for him, the issue is that people should have the right to access information about their own health. In this way, they can do their own risk assessment:

> Because if you just put this out and say, here is what the system is […] you make a decision about what you want to do just right now […] they make decisions if they want this product, if they don't want it, they can share information or not […] People already do it with 23andMe, there are so many companies out there providing direct information. This would do it ten times more. And I thought that it is good that people make their own assessment.

In line with this idea of the desirability of autonomous assessment of health, the role and importance of physicians in diagnostics was explored during discussion of scenario 2. One participant critically reacted to the idea of "depersonalization of medicine" articulated in the scenario, claiming that computers can do everything better and that there is no "particular value for personal contact". Not everyone agreed with this view. One participant explained that it depends on the type of disease and the exact situation: "some talk can be very useful". Another noticed that "patients do not know their symptoms for sure" and that "doctors can help understand". Later on, during discussion of phase 2 of this scenario, another participant pointed out:

> I think you have to be very careful about how the technology is introduced…they have the data, but to interpret it, I don't think should be the patient's task…that has huge implications for the patient…one has to stop and think about, how does that person respond to that

7.4 Techno-Moral Vignettes and Scenarios in Action 169

information? Doctors would rather not give you a diagostics rather than give you a wrong diagnosis, for example, if I suspect that the patient has prostate cancer, I will order for the test and I will wait until the test arrives before I say anything to the patient, because if the patient doesnpt have symptoms, you do not have the expectation that you have prostate cancer, not even remotely, so if you introduce this technology you also have to provide it with an interpretation....

This controversy amongst participants around the desirability of patient-centered versus doctor-mediated practice continued such that the discussion eventually moved on to the issue of "how much people really care about their health":

P1: this is a product and the first thing you have to do is to look for the market...we assume that our consumers are all people in US. How many people are willing to pay? How many people smoke? Or eat at mc donalds? The people who really care about their health on a regular basis are only a few and only those people, like 20 % are our consumers. And then you have to design the controversies only on those people....in that case you won't see the scenario 2.2
P2: unless part of the incentive for your insurance is that you participate in this test
P1: if you define the consumer, you know the mentality and you can address these issues

Participants agreed that "many people simply do not care", but while some of them thought that "You have to create the structure to motivate people to do it" (for example through an insurance system that benefits people who attend regular check-ups), others believed that a system like ImSg should just address "people who take vitamins".

This discussion opened up another topic: the relationship between information, empowerment and desirability:

P1: because more people have information, more widespread, the better...more knowledge about health
P2: me and you might do something with the information, but the other people...they want the doctor to tell them what they have to do...it depends on the individuals
P3: you think that information leads to better decisions but this is not necessary the case

Participant 3 explained this point further:

P3: because you empower people, it doesn't necessarily mean that is a good thing. If people have a big problem with patients who do not take care of themselves, and they go to the doctor...because ultimately somebody has got to pay for that, right? And so, you go to the doctor and it's expensive and somebody has to pay.... Because ...it costs... for somebody who takes responsibility to do it in an early stage, then this is part of the conflict, how you handle a situation like this....we pay those cases one way or the other, because people do not take responsibility...that's why I was wondering whether it would be possible to have some kind of virtual game to observe how people interact...just to see whether you can shape the debate, so that you can prepare people for this. Because the technology is gonna come one day, either it's immunosignature...or...I am wondering whether you can shape the attitude at an early stage...attitudes change really quickly.

The proposal of an online database is framed as a "sociological experiment". However, this experiment doesn't aim to learn about potential consumer demand, as other participants had mentioned through reference to marketing strategies. Instead, this experiment has the interventionist aim of shaping people's attitudes.

One participant also came up with a third scenario and a number of ethical controversies that they felt were missing in the two scenarios provided:

> When I was reading this stuff, I think the main issue of this technology in scenario one and scenario two that you mentioned...you always focused on the post-symptomatic, but I was expecting some controversy or some scenario 3 in which you actually state that, since we are moving back towards the pre-symptomatic, we are telling people that you might get the disease earlier, something like that. So XYZ [...] he or she finds out that she might be showing some part of the pancreatic cancer or some cancer signature so she freaked out and she went out and spent like 10,000 dollars and then later it turns out, you know, that it was just a little warning for the glute system [...] So I would see more that the controversy would occur in the early stage [...] if your dealing with cancer or diabetes stuff, there's a big window of 10 years and 5 years that you develop the cancer and there are so many early signs, so many early warnings there. At what point in time do you wanna decide what you wanna do. And at every time point there's a cost involved. So at what time point you can decide what you want to do and at what price [...] so i was thinking more over that would be a bigger controversy. [...] At what time should I seriously consider myself as a candidate for pancreatic cancer. You start going to tell me very early and it starts very late. and nobody knows the proper time. You don't want to freak me out telling me that i might and you don't want telling me that it's too late. But no one actually knows the proper time to tell a person. So that's, i think, the biggest hurdle there is, not this FDA approval of taking the misclassification, or errors or all this.

The Role of Scenarios How did the scenarios contribute to discussion of the desirability of Immunosignatures? As noted above, evaluation of these "grounded explorations" depends on their ability to broaden debate whilst still being considered plausible by discussants. Generally speaking, participants felt that the scenarios were extremely realistic and detailed. Although they didn't contest the plausibility of the scenarios, they often questioned their desirability. One participant was particularly critical of them:

> you know what bothers me about these scenarios? They take the whole technology and follow one event. The implication here is that we should make a decision on this one single event, even though in the context you cannot make this judgement, because there are other things...

As he explained later, "without numbers, without quantification it is difficult to make judgments. Here it is a psychological decision, it is very hard to do". Quantifications and numbers should be there in order to judge the desirability of an emerging technology. However, this is not about scenarios plausibility; instead, the participant is making a normative point. As he explained:

> unfortunately, it is probably a realistic scenario...that single events will have big influences and there is difficulty in having people doing a cost/benefit analysis...yes, there will be bad things happening, but there will be also good things happening.

What the participant is saying is that a quantitative cost/benefit analysis *should* guide decisions on the desirability of new science and technology, though *in fact* this is often not the case. This normative judgment is also evident when the timeline of the scenarios is commented upon:

> yes it might be that by 2017 we are going to be in trouble but I hope it is going to be faster than that. [...] I think it's too slow, I hope it's too slow.

7.4 Techno-Moral Vignettes and Scenarios in Action

Scenarios were often used by participants to articulate their own, normative views, and to discuss these with other participants. For example, one participant expressed their preference for scenario 1, in which data are available to all end users via websites, because he thinks that greater understanding and knowledge can be gained through this approach. Later in the discussion another participant said:

> P1: scenario 1 is my worst nightmare.
> P2: really? I liked it so much!
> P3: Scenario 1 is the Immunosignatures nightmare. Because yes, you enable people and you give them their information, but how you interpret is really really important.

This participant continued by explaining the ways in which the interpretation and mediation role of the doctor is important, particularly when dealing with healthy people who might not be ready to hear that a cancer is growing in them.

Commenting more explicitly on the outcomes of the discussion, participants agreed that interesting issues were tackled, because "when you present the scenarios everyone starts looking from different angles at what might happen". Although some of them (especially the junior researchers) added that "this doesn't change the way we do things…it is for later on", other perspectives emerged from the more senior researchers:

> Before I was more into the scenario 2 mindset, I thought that a diagnostic analysis (…) would be better, but now I think screw it, let's just make something that people can use, we can put the data out there and let's just see what happens….because there are very irrational ways to look at things that we as scientist can never anticipate. We never think, never anticipate what people will do with things…you can have an app store, or I pad, maybe a pharmaceutical company cannot make profit out of it, but….there might be more additional value that we cannot ancticipate, there is a lot of additional value that can come up from an information platform that we cannot anctipate.

Participants were not only willing to discuss the issues contained in the scenarios, but found them of interest, with the capacity to spark animated discussion. The normative content of the scenarios was discussed and reasons given to support or dismiss particular normative positions. Furthermore, as shown in the last quotation, discussion triggered by scenarios succeeded in broadening the considerations and evaluations of participants ("Before I was […] but now I think…").

It is important to note that the scenarios were not sufficient for broadening participant discussion. In fact, participants often tended to fall back into strategic thinking about how to make the technology more successful and to increase demand for it. In this workshop, reflection on the desirability of ImSg was often driven by a market logic. The role of the moderator was crucial in encouraging respondents to articulate and discuss the moral values implied by their arguments.

7.4.2 Workshop 2: the Nanopil

The second workshop was organized with the developers of the Nanopil (NP) technology. The workshop was organized over a year after the fieldwork described in Chaps. 3, 4 and 5, and took place on the day when the oncologist, the initiator of the NP, paid his monthly visit to the BIOS group at the University of Twente, the Netherlands. Besides the oncologist, all the researchers who were involved full time on the project participated in the workshop: the group therefore comprised one PhD student, two post-doctoral researchers, one assistant professor and one full professor. Furthermore, I invited another full professor who had acted as a speaker during a public debate on the NP. I had the role of chair, while a colleague of mine, aware of my agenda, intervened in the discussion whenever she felt that probing was needed. The workshop consisted of five rounds of about 20 min each. In the first, participants were asked to reflect on whether the NP is a good technology and on potential unwanted consequences. In each of the following three rounds a vignette was distributed and discussed. In the fifth round, participants were asked to evaluate the discussion.

Techno-Moral Vignettes Analysis of expectations of the desirability of the Nanopil suggested a number of existing (implicit) value conflicts ripe for exploration. I selected topics relevant to the NP developers so that they (could) engage them more into the discussion. I then composed short narratives in which the selected topics played out.

Vignette 1 is about a 22 year old girl who complains about having to take a laxative for the NP to work. This vignette, based on the considerations which emerged in Chap. 4 around current practice in colonoscopy investigations, compares the NP with the stool test (FOBT). The vignette was expected to provoke researchers by questioning their claim that NP is desirable because it is more user friendly than the stool test that is currently available. Its goal was to provoke discussion of whether user-friendliness was indeed built into the design of the NP and of the trade-offs embedded in that design.

Vignette 2 is about a man who is waiting for results from the pill whilst carrying on with his normal routine; however, anxiety and nervousness about the result distracts him so as to cause a small car accident. This vignette was based on the concerns of the general practitioner, presented in Chap. 5, about the desirability of self-testing. This vignette's goal was to instigate discussion of the desirability of NP's promise to allow users to test themselves anywhere at any time. The vignette also aimed to trigger participant reflection about the most desirable system for sending results to patients (blue bolus vs. wireless).

Vignette 3 is set in a doctor's office, where Mr Smit, who has been diagnosed with cancer, complains that when he did the screening using the NP 2 years earlier he didn't receive a result. The doctor observes that the assumption of the screening test is that people are responsible and autonomous: Mr Smit holds some responsibility for not taking action when he didn't receive the result of the test. Vignette 3 was

7.4 Techno-Moral Vignettes and Scenarios in Action

designed to trigger participant reflection about the fact that, though NP is expected to empower people, it also creates new worries, obligations and constraints: 'can' becomes 'ought'. In particular, it focuses on a shift in the responsibility attributed to patients for their health and in consequent increases in personal obligations. It aimed to foster participant discussion about the desirability of this broader social trend, to which they are contributing with the development of the NP. These three vignettes are fully attached in the Appendix of this chapter.

Opportunities for Exploration During the first round, participants were asked to articulate their expectations of the desirability of the NP. The voice of the oncologist was prominent in this part of the discussion. The topics were the same that emerged from the media sources and previous interviews: the suffering due to late discovery of cancer, the life-saving opportunities offered by screening techniques (exemplified in the cases of breast and cervical cancer), the burdens and unreliability of currently available screening techniques for colorectal cancer (the stool test), the high costs of colonoscopy-based screening, and the expected affordability and reliability of the NP. Participants were then asked to articulate potential unwanted consequences of this technology. This part of the discussion presented more heterogeneity. Issues that respondents mentioned included: potential fear of ingesting a piece of electronics; the possibility that users could trust the pill to the point that they neglect other signals of disease; privacy concerns because users might fear that the pill detects other types of information about the body and send it to interested parties; and, finally, false positives and false negatives.

In rounds 2–5 a broader range of topics was tackled. Throughout these rounds, the discussion often touched upon the theme of responsibility. Commenting on vignette 1, one participant observed that that:

> This person [in the vignette] doesn't want to face reality. The question is: Do you want to know? This is a difficult question. It's a matter of choice.

This comment opened up a discussion about whether people have the right to decide not to know about their health condition or if, instead, society should encourage them to test themselves. Reduced insurance premiums were proposed as a way that society could introduce an incentive for self-testing. A participant pointed out that this might not be desirable, and could in fact be considered another unintended consequence of the NP. This conflict between knowing and not knowing opened up the question of the information and education that would be provided to potential NP users. In the context of this discussion one participant noted that:

> Doctors should always explain to patients that they went through it [the test] themselves. If you have never experienced pain, if you don't know what pain is, you cannot tell to the patients…

So, doctors should try the NP themselves. Interestingly, the role of the doctor and her responsibilities turned to be quite relevant. In the fourth round, discussing the case in which the result didn't arrive to the patient, some participants agreed that responsibility should *never* be put on the patient. However, a participant reacted to this:

> I am not in favor of putting all the responsibility on the physician, I like if some responsibility is also on the patient. Because I see many cases in which the clinicians make mistakes and things eventually go right only when the family of the patients makes them notice it, or they ask them to look again.

Interestingly, another participant observed that in the case of a screening test, we should not talk about "patients" but "clients". Because they do not present symptoms, and consider themselves healthy, clients are less tolerant of undergoing complex procedures. For this reason, responsibility for the good quality of the screening procedure should not be put onto clients but onto the organizers of the screening program:

> It is also true that in general it is good to have everything, data information and files, under control. But in the case of the pill, you have to be perfect. Because you are asking people to come to you when they are healthy, you are asking them a lot and you cannot take any risk.

This question of responsibility in the case of a malfunctioning system brought the discussion back to the earlier point regarding the responsibility of the individual for testing her health condition, and the role of insurance companies in this:

> P1: The way I see it is: as insurance company client you have the choice of buying a package and if you go for one in which you pay less, then you accept to take responsibility.
> P2: Well I was wondering whether this is fair […]
> P1: […] this is their economic model.
> P3: But this is not a nice model…it is big brother like.

This unsettled controversy shows how the cases introduced by the vignettes did indeed spur debate on moral issues. Issues such as that of attribution of responsibility for individual health were largely discussed after the introduction of the vignettes. In this sense, the vignettes succeeded in putting new issues on the discussion table.

Besides opening up ethical controversies, participants were invited during the vignette rounds to address problems in a constructive fashion, such that discussion explored potential solutions to the problems raised in the vignettes. The NP was often acknowledged as being part of a broader techno-social system. This is shown in many of the responses participants gave, such as: "this is a matter of management of information", "this is a question of organization", "this is a problem of how the test is distributed", or "this is a broader (bigger) problem"; similarly, it was said to be "an issue of informing" or "educating". By emphasizing that good use of the NP depends on societal configurations, the role of design choices and development decisions (and consequent responsibilities for them) was played down. However, at some points "technical solutions" were pointed out. For example, while discussing the case of the user not receiving the results of the test, one participant underlined that in such a case the initiative to check that everything is going right "should be part of the organization". Another participant pointed out that the user might also not have taken the pill. The first participant responded with another type of solution:

7.4 Techno-Moral Vignettes and Scenarios in Action

> I think you can also have something on the pill that detects the temperature…if the pill is not at a body temperature the thermometer we see so…we can check whether the pill was taken.

In this way, the moral problem of ascribing responsibility for pill malfunction is partially addressed through a technological solution.

Finally, on several occasions participants reflected on the video that they use when they want to show a broader audience how the Nanopil works (see Chap. 3). At several points in the discussion, participants commented on the inadequacy of the video as a promotional tool. For example, during the discussion of vignette 2, one participant commented that:

> It is not the best video, because it doesn't give you an idea of the time dimension, it is very different from reality, it takes 10 or 20 hours before you receive the result […] this [idea] should be incorporated in the video somehow, maybe a little clock.

And later on:

> P1: For me the message is that it is benefial to have such a pill and that the education that we have done up to now is insufficient, it should be done better…the movie should be changed in various way to make it more clear to the clients. We got used to the movie, but now we know that many things do not work.
> P2: Also the doctor should be involved when making the movie, because now, the doctor was not involved…
> P3: Yes, I think it was a good approximation, at the time it was done, but we should fine tune it a bit more, but now we have to improve it and refine it.
> P2: I think it is also good to show that it allows a clean surgery
> P3: But of course a movie is always an approximation of reality, how much detail do you put in it?
> P2: It depends on who is your audience…it has to be such that the patient understands the background, because half of what you tell a patient they don't hear.

What is interesting in this exchange is that the participants, who have previously referred to the importance of good "education" and "information" on the Nanopil, reflect on the way in which they have been educating and informing. Participants thus re-evaluate the means that they have employed to communicate to a lay audience, and propose some strategies for improvement.

The Role of Vignettes Vignettes contributed to fostering discussion of the desirability of Nanopil in different ways. First, because of their thick narrative, the vignettes provided a concrete reference point during discussion. By recalling "this person" or "this case" in the vignette, participants contextualized their arguments, for example by saying "You might say 'what the hell, I am not doing it, I prefer to go the *movie*". Potentially abstract ethical questions are thereby tackled in a more concrete, contextualized way.

This detail rich character of the vignettes also has a further consequence: it makes some ethical concerns appear very situated, and therefore easy to dismiss – for example by suggesting that "this seems like a bad story of a bad laxative. It is not so bad to take the laxative". Similarly, the argument of vignette 1 can be disregarded as a function of the girl's personal attitude, rather than being addressed as a legitimate moral question. The narrative details of the vignettes, while contributing to the consistency of the story, therefore have the drawback of including variables that can be used by discussants to dismiss the argument by undermining its relevance. The plausibility of the vignettes and their grounding in concrete technical, social and moral aspects, which are acknowledged by stakeholders, here plays an important role. In fact, in order for discussion to proceed and the vignettes not to be considered farfetched, it may be necessary to explain that the "bad story of a bad laxative" is in fact the current story for patients who have to undergo a colonoscopy.

The vignettes, by proposing a snapshot of a problematic situation, were also found to trigger discussion of some useful topics not explicitly tackled by participants in the first round: for example, the burden on users, the anxiety of a portable/ingestible test, and where responsibility lies in the case of malfunction. Furthermore, the discussion also included other aspects not explicitly addressed in the vignettes, for example the quality of the video on the Nanopil. They therefore achieved the intended goal of broadening technology developers' discussions of the desirability of the Nanopil.

7.5 Discussion

This chapter has shown that the plausibility assessment not only provides an analytical tool for the ethics of emerging technologies, but that it also can support the creation of tools for intervening in the process of technology development. In particular, the plausibility assessment can help in building techno-ethical scenarios and techno-moral vignettes. In the context of interactive workshops, these scenarios and vignettes are a tool for triggering the moral imagination of participants and facilitating a grounded but exploratory discussion of the desirability of emerging technologies. With the goal of bringing ethical reflection "on the laboratory floor", these exercises aimed at specifying actual objects and practices as well as meanings and values related to research activities (Boenink 2013). Instead of coming from outside with an ethical assessment of the technology at stake, in the spirit of pragmatist ethics, these workshops aimed at articulating existing moralities in technology developer's discourses and practices and "integrating" (Fisher and Schuurbiers 2013) considerations on social and ethical aspects in the contextual practice of scientific and engineering research.

These tools were employed in two exercises that I designed with the goal of triggering the moral imaginations of technology developers around the desirability of the products of their research.[11] These exercises aimed to explore the use of vignettes and scenarios in such contexts, and to point out lessons for further use of these tools.

---

[11] An in-depth comparison between the two workshops should take into account the differences between visions of the Nanopil and the Immunosignatures, the disciplinary background of the participants, and cultural divergences in tackling ethical issues between the Netherlands and United States.

7.5 Discussion

In Sects. 7.4.1 and 7.4.2, I have shown that scenarios and vignettes were taken as a plausible point of departure and acted to broaden their discussions and considerations. I have also pointed out where there is room to improve these tools. For example, scenarios do not necessarily direct discussion towards issues of moral desirability as participants may be more focused on issues of profit, or economic desirability. As explained, a moderator can play a role in pointing out value conflicts in participant discourse. Similarly, narrative details and structures have the advantage of triggering participants' imaginations and making the stories more plausible. However, they may also confine a normative argument to a specific context.

These exercises have also contributed to the existing knowledge about the design of these kinds of tools. Firstly, in building vignettes and scenarios it is important that the elements of moral reflection should not be too numerous, in order to avoid discussion going in too many different directions. Because these elements of moral and ethical reflection are the outcome of a previous (plausibility) analysis, they may require more background and explanation than the narrative form can offer; the result therefore may be too complicated for readers who are not aware of the previous analysis. In this sense, vignettes provide a better tool because they provide only one or two issues for discussion. Second, the fact that I was well acquainted with the context, the participants, and their research allowed me to refer to specific cases and situations, not only in the vignettes and scenarios, but also during the discussion. For example, I could refer to specific projects or problems that I observed during my fieldwork. This helped to make the ethical discussion more concrete and relevant for participants. Third, scenarios and vignettes trigger the imaginative capacity of participants by thickening ethical considerations with concrete details. Details from interviews and fieldwork give a voice, sex, age, time and place to otherwise abstract and disembodied ethical arguments. These details, together with the consistency of the story, contribute to the plausibility of the scenario and discussants' suspension of disbelief.

Another point to be mentioned is that the two workshops were attended by relatively homogenous groups of stakeholders: other than the oncologist in the Nanopil workshop, the participants were all scientists/technology developers. As has been previously pointed out (Robinson 2010), heterogeneity is important in interactive workshops, especially in enabling the "heating up" of moral perspectives (Shelley-Egan 2011). The role of the oncologist in the Nanopil workshop was indeed important for bringing a different set of interests, values and principles to the discussion. However, the homogeneity of the group may also be functional; for example, it is logistically easier to gather a uniform group of participants, and this could encourage the use of such exercises on a more regular basis (especially at an early stage of technological development, when there is less at stake). It remains to be seen, however, how these scenarios could be used in different and less confined types of discussions in which several stakeholders are invited to deliberate.

A final reflection concerns the timing of this type of exercise. What is the right time for these exercises in moral imagination? ImSg and the NP are at an early stage of development. At this stage, technology developers feared that discussion of their desirability with a heterogeneous group of stakeholders would boost both hopes and hype. For this reason, the exercises were limited only to technology developers. Such exercises offer an opportunity to bring discussion of the desirability of emerging technologies into the laboratory, and to affect research decisions at this early phase. As one of the participants in the workshop on the Nanopil announced at the end of the session: "Next time, we will have the Nanopil lying on the table". However, it also emerged that some of the problematic situations introduced by the vignettes and scenarios were discarded by participants because they were considered to be too far along in the technology development process. In evaluating these exercises a graduate student commented: "I'm not sure whether it will actually affect any of my actions over the next year or two of my PhD research".

Although it is hard to track direct consequences for the development of the Nanopil and Immunosignatures, the two workshops I organized are a proof of principle that it is possible to create situated spaces to explore issues concerning the desirability of emerging technologies in a grounded manner. The plausibility assessment of expectations presented in this study plays a fundamental role in doing this. In fact, it points out implicit value conflicts in expectations of the artifact and between potential actors regarding its use. Furthermore, this assessment dismisses the instrumentalist misconception that a new technology will have predetermined, desirable consequences. In this way, the rich interactions between technologies, societies and morality are brought into upstream discussion of the desirability of emerging technologies. Embodying this analysis in a narrative form, techno-ethical scenarios and techno-moral vignettes are "grounded explorations" that aim at triggering the moral imaginations of their readers. On the one hand, they encourage discussion of emerging technologies that is not futuristic, because it is based on analysis of current practice. On the other hand, they also avoid foundationalism in their exploration of present decisions by assuming that our ideas of desirability and morality co-evolve with our social structures and technologies. The situated and local context of the workshop in which participants are involved in the development of a certain technology is both a limitation, as not all relevant stakeholders can have their voice, and a strength of a pragmatist approach that aims at privileging contextual assessments. The open question is of course to what extent such exercises can trigger moral imaginative discussions with a broader assemblage of stakeholders holding a more conflicting set of values.

A final remark: apparently, these scenarios fall in what Alfred Nordmann (2013) refers to as "plausibility^2" (plausibility squared), that is plausible scenarios describing consequences of emerging technologies in future worlds. As Normdann claims[12] such scenarios require to be assessed with respect to their internal plausibility as well as with respect to the assumption that they are plausibly describing a future state of affairs of the current world. This is very hard to assess since we don't have access to the future, and the only world we can know is the present one. According to Nordmann, we should be careful of using these "future" scenarios to make decisions in the present word. I believe it is very important to remark that although these scenarios are indeed a case of plausibility^2, they are not presented as tools for

---

[12] See also Chap. 2 for a discussion of the meaning of "plausibility".

decision-making, they do not provide plausible expectations that suggest action. Instead, the scenarios are plausible stories that aim at stirring moral imagination and encourage scientists and engineers to think about alternatives in current developments. What the scenarios do is to show that some design choices carry some moral consequences and invite a normative discussion on them. Such discussion is not about the future, but about the mundane current world.

Appendix: Techno-Ethical Scenarios and Techno-Moral Vignettes

Techno-Ethical Scenarios on Immunosignatures

Scenario 1 Part 1

Current Situation

Partly due to science and technology advancement, we are currently experiencing an increase in the longevity of the American population. This larger demographic of elderly persons, implies an increase in the amount of sick people, since are more prone to get sick or diseased. These costs are not sustainable in the long run.

A conspicuous number of experts from diverse fields points out that in order to address this problem, the adopted healthcare model has to move away from a symptomatic "one size fits all" treatment-based approach. A more affordable and effective model should lead towards (1) personalized and targeted care, (2) early and individualized diagnostics and (3) preventive risk predisposition.

These trends are visible in a broader landscape of actors operating within the sector of innovation in the American healthcare. To point out a few examples, hospitals promote innovative wellness programs in which patients are motivated to change their lifestyle; FDA encourages pharmaceutical companies to develop companion diagnostics; national and local policies of health educations raise citizens' awareness with prevention information campaigns.

The Biodesign Institute at Arizona State University is active in supporting science and technology contributing to this broader purpose. Immunosignature research, at the Centre for Innovations in Medicine, aims at developing a tool for early, pre-symptomatic and personalized detection of diseases. This preventive healthcare model promotes a **"patient-centered" approach**.

In a patient-centered vision of healthcare, people do not need doctors and are able to be in control of their health in an autonomous way. In the long term, this vision implies that people commit to a healthier lifestyle and are responsible for their health without the mediation of a doctor. Some online platforms like patientslikeme.com, curetogether.com, inspire.com are already heading in this direction: they create social networks in which patients can compare their data and support each other. The concept behind Immunosignaturing (IMS) is to contribute to this aim. It enables individuals to become more aware of their health and invites them to perform daily monitoring. With Immunosignaturing, as Dr Stephen Johnston explains, "each individual could see their own biosignatures and the implications of these signatures on a regular basis. This should empower them to take more responsibility for their own health and stimulate their own scientific education on what this means".

2012–2017

CIM directors decide to pursue the patient-centered vision and they contact different online platforms in which patients interact. In January 2013, the R&D director of the for-profit organization "patientslikeme.com", which provides a platform for patients to exchange information and "converts their stories in computable data", contacts the directors of the Centre for Innovation in Medicine. He has read about Immunosignaturing from their website (in which CIM recruits people who want to donate biological samples and share their health information to contribute to the IMS research). The two agree to sponsor the IMS project on the "patientslikeme" website in return for free immunosignatures for their users. In a video interview published on patientslikeme.com, CIM co-director Neal Woodbury explains the difference between IMS and genetic testing: "Genetic tests tell you about your genetic make-up, that you share with your family. Immunosignatures tell you a story that is unique to you. This is the story of the places you have visited, the food you have been eating, the things you have been doing. Your immune system stores all this information as a big book and by reading its pages day after day you can understand the implications of your daily choices on your health". A disclaimer on the website explains that IMS is still a developing technology and the data and its interpretation is still uncertain. After few months of collaboration, JameX, a diabetic user on "patientslikeme", compares his IMS with that of JulieD and comments on the blog about the surprising similarities between their condition and the IMS picture "These images are helpful beyond words… I feel connected to people who are suffering from the same symptoms as myself. And we know we can do something to change them. You have made me feel empowered…". Some smokers start an online support group on patientslike.me in which they motivate each other to quit smoking.

In September 2014, the behavioral psychologist Dr Shannon Kyle comments on the phenomenon of health social networks in Time magazine, "the IMS images are constantly changing and when patients see this they feel that they can control their health. Therefore they are motivated to change their smoking habits. The fact that they can openly share this information with other users commits them to a healthier behavior. Solidarity and community values play an important role in this". In February 2015 Google Health offers to have IMS has an available test on their website. Health Tell company, owned by Stephan Johnston and Neal Woodbury, starts doing IMS.

In December 2015, the Office of Privacy and Civil Liberties comments on the flow of sensitive data that endangers privacy. "Furthermore, information like that provided by IMS is not really robust. They show a picture with different colors that changes day after day, but what this means with respect to people's health is still unclear. Researchers are still working on this and the whole point of this website is to have free samples. It's just a business at the expenses of users, who should be protected". Some voices are raised by the community of patientslikeme.com against these accusations "People have the right to know about themselves, and IMS enables them to do so" "the whole idea of patientslikeme.com is to create a community in which people share stories and learn from each others, transforming their individual stories into a learning experience for others like them. IMS simply works in making people aware of the consequences of their lifestyle. Even if the meaning of the information is not clear yet, it does something good for people. And this is what matters".

In April 2016, Sally Meine, a 19 year old girl from Ohio, is admitted to the Cleveland Clinic in a coma, and is diagnosed with extreme malnutrition. When she regains consciousness, she explains to the doctors that she was experimenting with her friends to see how her immunosignature would change if she ate only fruit. The parents of Sally sue patientslike.me and Health Tell for liability for not putting in place enough failsafe to prevent misuse of IMS technology. In May, Sally's case is brought up in FDA discussions about the large availability to a broader public of sensitive medical data that is still not reliable.

In December 2016, the CEO of a famous pharmaceutical company criticizes the FDA intervention. In an editorial, he highlights that collaboration between HealthTell and patientslikeme.com are promising because they enable profiling of a large number of people with high degree of statistical relevance, advancing risk assessment and steering the direction of company research: "these experiments are nothing new, they happen everyday with Google. Why shouldn't we use the same tools to develop our knowledge of health conditions and ultimately make our healthcare system more effective?"

The FDA's objection on the disclosing of Immunosignatures to patients on online platforms is legally reinforced due to the "current lack of reliability" of the information provided. However, an increasing consensus in industry and political environment acknowledges the importance that research on personalized signatures could have for national healthcare. Following a political controversy on the desirability and feasibility, in November 2017, policy agreements are made:patients' medical histories and biological samples analysis will be stored online. This information is openly available to researchers, while maintaining patient anonymity. Educational campaigns are promoted to encourage people to contribute to this database: "Do it for your future. Do it for your children. Do it for a world without patients".

Scenario 1 Part 2

2018–2024

This decision boosts the development of IMS analysis. In 2020, a normalized "healthy" signature is established (including an algorithm to account for individual variation), leading to an improved detection and determination of illness. In 2022, local governments start some wellness programs, educating citizens to adopt a healthier behavior. IMS is proposed as a good tool for educating healthy citizens to take more responsibility for their health.

In January 2023, in the state of California programs are initiated to provide people with lower incomes access to IMS at home. This is undertaken as an experiment to see if they engage in a healthier lifestyle. To make it more compelling, people are offered public Medicare in exchange for monitoring themselves with IMS on a weekly basis. Eligible citizens have to pick up the test at the closest Urgent Care clinic and send a biological sample to be tested with IMS. They receive the results via email and they can check their medical record online. They are told if they should inform the Urgent Care in case of serious and continuous abnormal signature.

In August 2026, the case of Francis Caine raises the public interest. In June 2024, Francis had received a message that she had to go to the urgent care clinic, but she had failed to do so. After 1 year she was diagnosed with cancer. The hospital had contacted the governmental Medicaid and they discovered that she was in the program but she had never acted on it. After the surgery, she is required to pay, because she didn't contact the urgent care immediately. In an interview Francis explains: "I didn't have any symptom back then, I felt good, I just did not think that it could be serious". Her neighbor comments, "She was already lucky not to pay for the medical insurance, but she did not care that the government and other citizens are paying for her. She was given the chance of being proactive, but she decided not to…now she has to pay". However, patient advocates take the side of Francis: "Discovering that something is wrong with you, that you might have a disease, that your life is threatened is overwhelming. We cannot expect that a single individual can cope with this information alone and simply act as others have told her to act. With this IMS idea, institutions are unloading their responsibility onto the patients rather than providing them with guidance and care".

Scenario 2 Part 1

Current Situation

Partly due to science and technology advancement, we are currently experiencing an increase in the longevity of the American population. This larger demographic of elderly persons implies an increase in healthcare costs, since the elderly are more prone to sickness or diseased. These costs are not sustainable in the long run.

A conspicuous number of experts from diverse fields points out that in order to address this problem, the adopted healthcare model has to move away from a symptomatic "one size fits all" treatment-based approach. A more affordable and effective model should lead towards (1) personalized and targeted care, (2) early and individualized diagnostics and (3) preventive risk predisposition.

The Biodesign Institute at Arizona State University is active in supporting science and technology contributing to this broader purpose. Immunosignature research, at the Centre for Innovations in Medicine, aims at developing a tool for early, pre-symptomatic detection of diseases. This preventive healthcare model promotes a **personalized diagnostics:** "you are compared with yourself yesterday". This long-term vision of pre-symptomatic and personalized diagnostics is expected to go through an earlier phase of personalized (symptomatic) early diagnostics. The application of IMS in a clinical context presupposes a system in which doctors play a central role and in which technology is expected to offer a tool for them to improve their practice. In this sense, IMS is considered as a tool to improve the current diagnostic practice.

Ongoing research at CIM focuses on screening and diagnostics for chronic diseases (i.e. Alzheimer), monitoring for cancer recurrence, and diagnostics for infectious diseases like Valley Fever. The potential diagnostic tools arising from this research are expected to be used in a clinical context.

Half of those affected by Valley Fever in the U.S. are in Maricopa County. Two-thirds are in Arizona. Fungus in the soil, especially during the dry, windy season, becomes airborne and settles in the lungs to cause the disease. In 2006 Tucson-based C-Path also helped UA scientists receive Orphan Drug Status from the U.S. Food and Drug Administration, which would allow scientists to apply for grants to help pay for the development of the drug.[13]

In 2011, coccidioidomycosis (Valley Fever) is reported by the Council of State and Territorial Epidemiologists (CSTE) as a nationally notifiable disease. This increases the publicity and mindshare in the public health community.

2012–2017

In June 2012, a study of the Centre for Disease Control highlights the healthcare costs of Valley Fever treatment for the State of California. In this study, the costs of absenteeism and presentism in the workplace due to Valley Fever are emphasized. At the end of 2012, Valley Fever Therapies LLC forms a partnership with the state of Arizona to take the drug to the market. "The state of Arizona thinks that this is an important enough public health issue" says Dr Galgiani, director of the University of Arizona Valley Fever Center for Excellence. Under these circumstances a diagnostics for Valley Fever becomes critical. Dr Galgiani and Dr Johnston, the CIM co-director and Health Tell CEO, investigate Clinical Laboratory Improvement Amendments (CLIA) certification for an immunosignature test.

---

[13] http://www.bizjournals.com/phoenix/stories/2006/05/01/story4.html?page=all

In December 2013, the American Medical Association (AMA) includes the Immunosignature test for Valley Fever in Current Procedural Terminology in Category III.[14] As a member of the board of trustees of the American Medical Association explains: "Primary care physicians in Arizona have been encouraged to enroll in educational programs about early diagnosis of Valley Fever. AMA considers this of utmost importance in the Valley area in order to improve patient care by reducing patient anxiety and unwarranted use of antibacterial agents. Moreover, serious complications requiring treatment might be identified sooner."

In May 2014, the state of Arizona initiates a public campaign to raise people's awareness of symptoms and potential treatments for Valley Fever. "My friend Karen died of complications from Valley Fever while she was pregnant. She was not diagnosed in time. This was at the beginning of 2000" explains Mia Valdivia, patient advocate for Valley Fever Survivors. "Since then, many things have changed. In this last year, I have received so many letters from people who go to primary care with few symptoms, are diagnosed with Valley Fever within one week, are quickly treated and keep on living normal lives".

In December 2015, Mr Carl Carlson was admitted to the Mayo Clinic in Scottsdale with advanced lung cancer. "He was diagnosed with Valley Fever one year ago with one of those tests. The doctor prescribed him some drugs, but the symptoms never left. Now they say that for all this time the cancer was growing. I want to understand who is responsible for this case of malpractice, the laboratory or the doctor?" says his wife, interviewed by a local radio channel. "The fact is that these immunosignature tests are optimized on some statistical data, the physician should examine the patient carefully and provide a more personalized diagnostics" the PR of a diagnostic company explains.

Following up this case, the National Center for Ethics in Health Care (NCEHC) issues a report on the use of laboratory diagnostics in primary care practices and emphasizes the importance of traditional diagnostics. "Primary care medical clinics in Arizona are financially benefiting from the introduction of IMS diagnostics in the CPT. For each of these tests they get a service fee, but they do not take time to actually visiting the patient, ask about the symptoms, provide high quality care and information. But of course these tests are never 100 % reliable. They should spend more time looking at the personal history and situation of the individual patient".

In April 2016, The association of American Physicians responds to this report explaining the worsening of their working conditions "primary care doctors earn

---

[14] Current Procedural Terminology (CPT®) is a listing of descriptive terms and identifying codes for reporting medical services and procedures. The purpose of CPT is to provide a uniform language that accurately describes medical, surgical, and diagnostic services, and thereby serves as an effective means for reliable nationwide communication among physicians, and other healthcare providers, patients, and third parties. CPT is also used for administrative management purposes such as claims processing and developing guidelines for medical care review. The assignment of a CPT Category III code to a service does not indicate that it is experimental or of limited utility, but only that the service or technology is new and is being tracked for data collection. http://www.ama-assn.org/ama/pub/physician-resources/solutions-managing-your-practice/coding-billing-insurance/cpt/cpt-process-faq/code-becomes-cpt.page

much less than other specialists and have to work in more difficult conditions. We can afford just 20 min per patients. How do you think we can provide good quality care? Furthermore, the spread of new diagnostic tests and new procedure demand a lot of preparation and retraining. Often these tests do not give a yes/no answer, but a grey result that needs to be interpreted. This is the responsibility of the doctor. If we spend all of our time learning about the possibilities for personalized diagnostics, when can we actually provide more personal care?".

In January 2017, the Centre for Medicare and Medicaid discusses the possibility of moving the Valley Fever diagnostic procedure to retail clinics (located inside retail/grocery stores) and staffed by non-physician providers, such as physician assistants and nurse practitioners, with remote physician supervision. This would release primary care doctors from the burden and decrease the service fees. However, due to its serious consequences the AMA points out that Valley Fever cannot be considered as a minor illness and should therefore be treated by physicians.

Scenario 2 Part 2

2017–2025

In 2017, there is a breakthrough in the collaboration between CIM and Mayo clinic: statistically relevant differences between immunosignatures of esophagus cancer patients and patients whose esophagus cancer has been eradicated are established. This enables researchers to optimize a test to monitor the recurrence of the cancer.

In 2020, the test receives CLIA certification, however the previous controversies rose with the valley Fever makes the AMA wary about inserting this test into the CPT lists. "This test opens the possibility for a cheap monitoring of cancer patients, but who should read these signatures? The oncologist or the GP? It is too delicate for GP, but the benefit of the test is to alleviate the work of oncologists." The test does not take on with GP's, but is taken up and appreciated by oncologists.

In 2023, Cleveland Clinic is interested in using the immunosignaturing test for cancer recurrence in their "health care at home" program. This program was initiated by the innovation department of the clinic as an attempt to respond, on one hand, to pressures of private insurance companies for reducing patients' visits to the outpatient clinic, and on the other hand to the pressure for a more continuous and effective connection between the patients and the care provider. With this program of remote healthcare patients are connected to the clinic via online facilities on a daily basis and the costs of visits to the clinic are kept lower. IMS is introduced as a home test for cancer patients to monitor eventual recurrencies. Every three months patients are sent at home a kit from the hospital and asked to send a biological sample by mail.

The head of the innovation department of Cleveland Clinic explains that this program makes healthcare a continuous and daily process, rather than an episodic one, "We offer a better quality care". "This is the future of healthcare, continuous monitoring of vitals and other values". You don't need to go to the hospital anymore. Payers are satisfied for the reduction of the costs.

However, George Carter, a 58 year old cancer patient, is found dead at home by a friend. In a letter he describes how he knew that is tumor was growing, but he was too afraid to go through the surgery again. He stopped collecting samples. "He was missing the personal relation with his doctor, when you don't have a person on the other side who is taking care of you, what do you do? Do you really call it personal care? Where is the person here?". The hospital responds that results of the test were returned to the patient with an advice to contact them, but were not actively followed up by the hospital. "We put the patient autonomy on the first line. It is our core value to respect the decision of our patient, whether they decide to go for a treatment or they decide to decline it."

Techno-Moral Vignettes on Nanopil

Vignette 1

"I am not gonna do this again, it's disgusting!"

"Listen, Nya, I'm tired of this. Try to behave like an adult, you are 22 now! You know why you have to drink this laxative."

"Yes, for the stupid pill to work....."

"This 'stupid' pill is an easy way to check that everything is fine. Your dad's family has a history of Colorectal cancer so you had to start screening early. Consider yourself lucky, 20 years ago people had to collect a sample of their stool, smear it on a sample card, compile it with their information, seal it and mail it to the lab. The pill makes this much more simple, comfortable and clean!"

"SIMPLE, COMFORTABLE, and CLEAN???? Why don't YOU try drinking this crap? And this unbearable nausea. Blech. I feel like I have to throw up after every sip. Having to run to the toilet every half an hour is clean? Joyce wanted to go to the cinema with me, but I can't! I have to be at home, drinking laxative, feeling sick and running to the washroom every 10 minutes. I feel like I am spending the whole day in the bathroom. I would rather spend 1 minute collecting samples and forgetting about it. But instead, I have 2 more liters of laxative to go. ARGH..."

"Hun, you are behaving like your grandma! Just drink it, the doctor said..."

"I don't care about the doctor, I am not gonna drink it all."

"And if the pill isn't going to work then?"

"Even better, then they will think that I am fine and they will leave me alone."

Vignette 2

Policemen: "Your identification and car documents, please."

Mr Watson: "Oh sure, I am really sorry, I am still in shock. The car is really damaged!"

Policemen: "Can you explain what happened?"

Mr Watson: "Yes, but I guess I have to explain it from the beginning. As every Wednesday, I take time off work to take care of the children. But today I was also supposed to screen for colorectal cancer. You know, I am over 50 and they sent me a test for checking if I have cancer. It is one of those pills that you swallow and it checks your intestine from inside and sends a report of the test to your mobile phone and to the doctor. Anyways, I was really eager to take this test, because lately I experienced some pain in the lower abdomen and I feared that there could be something wrong. The fact is, I was very nervous about the result of the test. So after swallowing the pill I waited and waited and I couldn't accomplish much besides imagining the pill floating within my intestine. But the message hadn't arrived yet, when it was time to leave home and pick up the children from school. So I went."

Policemen: "Are you allowed to drive when you take this pill?"

Mr Watson: "Well, it is said that with the pill you can test your health everywhere and once you have the signal-receiving belt on and your mobile phone close to you, it's just fine. However, while driving, I felt again a stinging pain in my abdomen, so when I felt the vibration of my mobile phone finally notifying me about my health status, I was distracted from observing the street. So I didn't see that the car in front of me was so close. I am terribly sorry, but how can you focus on the street, when you are afraid that your health is at stake?"

Vignette 3

Doctor Jansen smiled while the next patient walked through his office's door. This should be Mr Smit. He opened Mr Smit's personal record on his computer and looked at it. Mr Smit was there because the result of colonoscopy was positive and a neoplasm had been found. Dr Jansen told Mr Smit about the advanced status of the cancer and the need for an immediate intervention.

"I don't understand, doctor, 2 years ago I participated in the national screening program and I took a pill to check whether everything was fine…"

"Yes, actually, I can see it on your record, but there is no result associated, are you sure you did the test?"

"Yes, I am, I remember it quite vividly because I hated having to take this laxative! I don't think I heard anything about the test afterwards, but I simply assumed that it meant I was OK. So how is it possible that this tumor is so big already? It must have been there two years ago! It would have been smaller then. What's the point of a screening that doesn't work?"

Doctor Jansen wondered, what had gone wrong two years ago? The pill, or the network? Maybe Mr Smit had done something wrong? Or maybe he himself had done something wrong. But with all those screening tests around, how to keep track?

"Mr Smit, please calm down, what did you do when you didn't receive a message? Did you call here to know more? That would have helped in tracking the problem at the time."

"Doctor, how could I have known? I took the test, nobody contacted me, so I thought I was fine, I thought I was under control. That's what I think when I am in a screening program, when I perform such an innovative test, sent to me by this screening organization. I think that everything is under control."

"I am sorry Mr Smit, but I must correct you. This screening program, like other do-it-yourself-prevention programs, presupposes that you are a responsible and autonomous person, who can take care of monitoring his own health condition. But this also means that you are expected to be more proactive. You have been negligent, Mr Smit."

References

Appiah, Kwame Anthony. 2010. *Cosmopolitanism: Ethics in a world of strangers (issues of our time)*. New York: W. W. Norton.
Ayer, A.J. 1968. *The origins of pragmatism: Studies in the philosophy of Charles Sanders Pierce and William James*. San Francisco: Freeman, Cooper and Company.
Boenink, Marianne. 2013. The multiple practices of doing'ethics in the lab. In *Ethics on the laboratory floor*, ed. Simone van der Burg and Tsjalling Swierstra. Basingstoke: Palgrave Macmillan.
Boenink, M., T. Swierstra, and D. Stemerding. 2010. Anticipating the interaction between technology and morality: A scenario study of experimenting with humans in bionanotechnology. *Studies in Ethics, Law, and Technology* 4: 1.
Brom, Frans, and Rinie van Est. 2011. *Technology assessment as an analytic and democratic practice*, Encyclopedia of applied ethics, 2nd ed. Amsterdam: Elsevier.
Brown, N., B. Rappert, A. Webster, C. Cabello, L. Sanz-Menendez, F. Merkx, and B. Van der Meulen. 2001. *Final report of the FORMAKIN project*. SATSU: University of York.
Burg, S. 2010. Ethical imagination: Broadening laboratory deliberations. In *Emotions and risky technologies*, ed. S. Roeser, 139–155. Dordrecht: Springer.
Coeckelbergh, M. 2007. *Imagination and principles: An essay on the role of imagination in moral reasoning*. Basingstoke/New York: Palgrave Macmillan.
Dewey, J. 1922. *Human nature and conduct; an introduction to social psychology*. New York: Holt.
Dewey, J. 1929. *The quest for certainty: A study of the relation of knowledge and action*. New York: Minton, Balch.
De Vries, G. 2002. Pragmatism for medical ethics. In *Pragmatist ethics for a technological culture SE - 12*, vol. 3, ed. Keulartz Jozef, Korthals Michiel, Schermer Maartje, and Swierstra Tsjalling, 151–164. Dordrecht: Springer.
Fesmire, S. 2003. *John Dewey and moral imagination: Pragmatism in ethics*. Bloomington: Indiana University Press.
Fisher, E. 2007. Ethnographic invention: Probing the capacity of laboratory decisions. *NanoEthics* 1(2): 155–165.

References

Fisher, Erik, and Daan Schuurbiers. 2013. Socio-technical integration research: Collaborative inquiry at the midstream of research and development. In *Early engagement and new technologies: Opening up the laboratory SE – 5*, Philosophy of engineering and technology, vol. 16, ed. Neelke Doorn, Daan Schuurbiers, Ibo van de Poel, and Michael E. Gorman, 97–110. Dordrecht: Springer.

Grunwald, Armin. 2004. The normative basis of (health) technology assessment and the role of ethical expertise. *Poiesis & Praxis: International Journal of Technology Assessment and Ethics of Science* 2(2–3): 175–193.

Habermas, J. 1973. Wahrheitstheorien. In *Wirklichkeit und Reflexion*, ed. H. Fahrenbach, 211–265. Pfullingen: Walther Schulz zum sechzigsten Geburtstag.

Habermas, J. 1990. *Moral consciousness and communicative action*. Cambridge, MA: MIT Press.

James, W. 1956. *The will to believe and other essays in popular philosophy, and human immortality. [1897]*. New York: Dover Publications.

Keulartz, J., M. Schermer, M. Korthals, and T. Swierstra (eds.). 2002. *Pragmatist ethics for a technological culture*. Deventer: Kluwer Academic Publishers.

Krabbenborg, Lotte. 2013. Dramatic rehearsal on the societal embedding of the lithium chip. In *Ethics on the laboratory floor*, ed. Simone van der Burg and Tsjalling Swierstra. Basingstoke: Palgrave Macmillan.

McGee, G. 2002. Pragmatic epistemology and the activity of bioethics. In *Pragmatist ethics for a technological culture SE - 8*, vol. 3, ed. Keulartz Jozef, Korthals Michiel, Schermer Maartje, and Swierstra Tsjalling, 105–117. Dordrecht: Springer.

Murphy, J.P. 1990. *Pragmatism: From Peirce to Davidson*. Boulder: Westview Press.

Nordmann, Alfred. 2013. (Im)Plausibility2. *International Journal of Foresight and Innovation Policy* 9(2–3–4): 125–132.

Peirce, C.S. 1960a. The fixation of belief [1877] 5.358–387. In *Collected papers of Charles Sanders Peirce*, ed. C. Hartshorne and P. Weiss. Cambridge: Belknap Press of Harvard University Press.

Peirce, C.S. 1960b. What pragmatism is [1905] 5.411–437. In *Collected papers of Charles Sanders Peirce*, ed. C. Hartshorne and P. Weiss. Cambridge: Belknap Press of Harvard University Press.

Putnam, H. 1987. *The many faces of realism*. La Salle: Open Court.

Rip, Arie, und Haico te Kulve. 2008. Constructive technology assessment and socio-technical scenarios. *Nanotechnology* 1: 49–70. doi:10.1007/978-1-4020-8416-4_4

Robinson, D.K.R. 2010. *Constructive technology assessment of emerging nanotechnologies experiments in interactions*. Enschede: Proefschrift Universiteit Twente.

Rorty, R. 1982. *Consequences of pragmatism: Essays, 1972–1980*. Minneapolis: University of Minnesota Press.

Rorty, R. 1989. *Contingency, irony, and solidarity*. Cambridge: Cambridge University Press.

Schot, J., and A. Rip. 1997. The past and future of constructive technology assessment. *Technological Forecasting and Social Change* 54(2–3): 251–268.

Shelley Egan, C. 2011. *Ethics in practice: Responding to an evolving problematic situation of nanotechnology in society*. Enschede: Proefschrift Universiteit Twente.

Stemerding, D., T. Swierstra, and M. Boenink. 2010. Exploring the interaction between technology and morality in the field of genetic susceptibility testing: A scenario study. *Futures* 42(10): 1133–1145.

Swierstra, T., and H. te Molder. 2012. Risk and soft impacts. In *Handbook of risk theory*, ed. S. Roeser, R. Hillerbrand, M. Peterson, and P. Sandin, 1050–1066. Dordrecht: Springer.

Swierstra, T., R. van Est, and M. Boenink. 2009a. Taking care of the symbolic order. How converging technologies challenge our concepts. *NanoEthics* 3(3): 269–280. Springer Netherlands.

Swierstra, T., D. Stemerding, and M. Boenink. 2009b. Exploring techno-moral change: The case of the obesity pill. In *Evaluating new technologies*, vol. 3, ed. P. Sollie and M. Düwell, 119–138. Dordrecht: Springer.

Thompson, P.B. 2002. Pragmatism, discourse ethics and occasional philosophy. In *Pragmatist ethics for a technological culture*, ed. J. Keulartz, M. Schermer, M. Korthals, and T. Swierstra. Deventer: Kluwer Academic.
van der Burg, S. 2009. Imagining the future of photoacoustic mammography. *Science and Engineering Ethics* 15(1): 97–110.
van Merkerk, R., and R. Smits. 2008. Tailoring CTA for emerging technologies. *Technological Forecasting and Social Change* 75(3): 312–333.
van Notten, P.W.F., et al. 2003. An updated scenario typology. *Futures* 35(5): 423–443.
Von Schomberg, René, Ângela Guimarães Pereira, and Silvio O. Funtowicz. 2005. *Deliberating foresight knowledge for policy and foresight knowledge assessment*. A working document from the European Commission Services, Luxembourg.
Zwart, H. 2002. Philosophical tools and technical solutions. In *Pragmatist ethics for a technological culture*, ed. J. Keulartz, M. Schermer, M. Korthals, and T. Swierstra. Deventer: Kluwer Academic.

Chapter 8
Building-Blocks for Ethical Assessments of Emerging Technologies

> *The subjects of our deliberation are such as seem to present us with alternative possibilities: about things that could not have been, and cannot now or in the future be, other than they are, nobody who takes them to be of this nature wastes his time in deliberation. (Aristotle- Rhetoric [1357a])*

Abstract In summarizing the main contributions of the book, this last chapter examines in what ways the approach discussed so far addresses the question, articulated in Chap. 1, of how to integrate normative sensitiveness in TA. It demonstrates in what way addressing the question of plausibility allows ethicists (as well as social scientists, bioethical committees, policy makers, technology developers, etc.) to address the normative questions that surround the issue of social desirability of emerging technologies with a broader set of visions wherein the material morality of artifacts, the worldviews of stakeholders and the dynamics of moral changes are spelled out. The chapter also considers the role of the "ethicist" in such processes and concludes by outlining some open questions for further research.

Keywords Deliberative democracy • Plausibility assessment • Pragmatist sites for normative analysis • Ethics of promising

8.1 Between "Grounding" and "Exploring"

The question of *how to engage in a prospective reflection on the desirability of emerging technologies* has been addressed in the policy domain within the Technology Assessment tradition, as well as by academics involved in studies of Ethical Legal and Social Aspects (ELSA) of new science and technology. As discussed in Chap. 1, assessments of emerging technologies retain an intrinsic normative connotation, in the sense that they evaluate technologies that are still under

---

In order to avoid repetition, this conclusive chapter summarizes the main points of this book and offers final remarks, intentionally omitting references to the literature that has already been discussed in previous chapters. The reader interested in a discussion of the literature should consult Chaps. 1 and 2 of this book.

development based on particular norms, values and ideas of "good". Second-generation TA approaches have explored methodologies that encourage a broad range of stakeholders to engage in focused discussions around the issues of emerging technologies that go beyond the "economic" to include social aspects. These approaches, however, do not seem to engage in a systematic exploration of the moral aspects of emerging technologies as well as in normative assessments. This types of assessments have been conducted by applied ethicists whose analyses, however, have been critiqued for being either too uncritical towards the feasibility of technological developments or for not taking into account socio-technical dynamics. Since technologies, people and values do not sit in a societal vacuum, it is crucial to take the societal context on board when assessing the desirability of technologies. Furthermore, an assessment based on a specific normative framework requires prioritizing some values over others. It also fails to recognize that a plurality of perspectives conflicts with the philosophy of the so-called "second-generation TA", which turns science and technology policy decision-making into a more participatory and democratic exercise. The question that emerged in Chap. 1, therefore, was: how to integrate a normative appraisal in democratic exercises of technology assessments that accounts of the complexities and diversity among the involved stakeholders and the societal context?

Within this general context one specific issue emerges as needing attention: the prospective character of emerging technologies. The issue is how to guarantee an exploration of their desirability – the good that is associated with them – without falling into the trap of producing a speculative normative analysis around some hypothetical technology in a fictitious future world? How to stay away from imagining implausible consequences and evaluating them on the basis of a fixed and socially unaware normative framework or theory? Technologies that are still emerging raise some epistemological challenges for those who want to engage in their assessment. Such challenges involving "future" thinking are even more daunting when we want to question the very "goodness" of a technology, which is in itself a "soft", unquantifiable criterion that people in a liberal society will likely disagree on (Swierstra and te Molder 2012). At the early stages of technological development, "expectations" and "visions" are the starting points of any technology assessment. Disseminated by technology developers, industry, media and policy-makers, expectations surrounding emerging technologies are uncertain, strategic and embedded with normative connotations. Ethics of emerging technologies seem therefore to stand between, on the one hand, the need to avoid excessive speculation and being *grounded* in the "here and now" (Nordmann 2007; Nordmann and Rip 2009), and on the other hand, the demand for *exploration* of the meanings and implications of current promises (Grunwald 2010).

The aim of this book has been to find a balance between "grounding" and "exploring" the desirability of emerging technologies, by discussing a preliminary analysis of expectations' *plausibility* (Chap. 2). Such an analysis disentangles and assesses the likelihood of statements that claim a technology will realize certain desirable worlds. Presented as an analytical and methodological framework, the plausibility assessment consists of an analysis of expectations and visions that the material

8.1 Between "Grounding" and "Exploring"

artifact will operate in a certain way (Chap. 3) and that it will be *used* in specific ways in dedicated social contexts and by particular actors (Chap. 4). These analyses offered grounds for assessing the plausibility of the expectation that a technology will have a social *value* (Chap. 5). Through analyzing expectations surrounding the Nanopil and Immunosignatures (Chap. 6), the theoretical and methodological aspects of this framework and the types of analysis it enables, were explained.

In Chap. 7, I discussed the pragmatist ethics approach as a viable normative framework for technology assessment because of its sensitiveness to different moral contexts and an embracement of the normative ideal of deliberative democracy. Pragmatist ethics provides a normative framework that encounters the sociological, contextual, process oriented goals of some forms of TA. Its general aim is to improve the process of public deliberations by improving its conditions in three respects: (1) including different perspectives in the deliberative process; (2) articulating the reasons, meanings and assumptions behind a problem; and (3) exploring possible scenarios of how new technologies change our moral concepts and vocabularies. In this context, the role of the ethicist consists of "facilitating public debate and political decision-making on emergent moral problems" rather than offering solutions (Keulartz et al. 2002: 15). In this sense, it is based on a "formal" approach of developing procedures that allow all interested parties to have a chance to actively participate in ethical debates. At the same time, this approach also aims at improving the quality of such debates with more substantive interventions that focus on the quality of opinions and interests. One way of achieving this is by triggering stakeholders' moral imagination around certain situations related to emerging technologies. Chapter 7 then described and discussed two exercises designed to engage technology developers in discussions on the desirability of their object of research with the use of two tools to trigger their moral imagination (techno-ethical scenarios and techno-moral vignettes), which had been developed on the basis of the plausibility assessment.

Consequently, this study aims at offering normative ethicists a methodological and analytical framework designed to address questions concerning the plausibility of emerging technologies. By analyzing expectations surrounding the emerging artifacts and their context of use, ethicists can make their normative assessments of new and emerging technologies more grounded, and therefore less speculative. Such an analysis also addresses the "normative deficit" of Technology Assessment. This analysis offers a tool that can design exercises of democratic deliberations that explicitly address moral issues related to emerging technologies. The plausibility approach achieves these aims by: eliciting the beliefs and values that are inscribed in the materiality of an artifact; pointing out the interests and normative positions of actors involved in the expected context of use; and, accounting for the non-linear impacts of emerging technologies. Articulating the moral values and norms in expectations of emerging technologies can be considered the first step towards their assessment. In line with a pragmatist approach, the plausibility analysis enables "occasional" (Thompson 2002) and plural normative assessments in which problems are considered as contextual – dependent on time and space – rather than fundamental and radical, and as such, conclusions are also perceived as contextual and provisional.

8.2 Towards Ethical Assessments of Emerging Technologies

I argued that the analysis of emerging technologies' plausibility is the first step towards an ethical assessment of emerging technologies. How is it so? Addressing the question of plausibility allows ethicists (as well as social scientists, bioethical committees, policy makers, technology developers, etc.) to pose normative questions surrounding the social desirability of emerging technologies with a broader set of visions wherein the material morality of artifacts, the worldviews of stakeholders and the dynamics of moral changes are spelled out. The goal of such ethical assessments is not to offer a punctual normative answer regarding some emerging technologies, but to improve the process of deliberation by enriching the quality of the discussion around their desirability. Therefore, the ideal of democratic deliberation does not only entail the inclusion of everyone who will be affected by a technology in the assessment of its desirability. Such an ideal also entails an improvement of the quality of the discourse on emerging technologies. The quality of the discourse is not improved in abstract or idealized deliberations in which a community exchanges judgments and reasons for decisions. Instead, normative assessments are localized and therefore happen within communities of stakeholders, in concrete contexts where strategic considerations, interests, and negotiations among stakeholders take place and in which the different normative perspectives should be disentangled. The plausibility analysis proposed serves the goals of ethical assessments of emerging technologies in different ways: it reduces speculation about the desirability of emerging technologies, helps in defining problems, highlights hidden assumptions, disentangles normative perspectives, avoids polarizations, describes practices and meanings expanding current moral vocabularies and finally, it enriches scenarios. Let us consider these aspects one by one.

First, it contributes to a reduction in speculations about the desirability of implausible contexts. For example, it was discovered that the Nanopil cannot be stored for a long time since the DNA molecules that are attached to the lab-on-a-chip within the pill have a very short lifespan (see Sect. 3.5). This means that controversies on whether the Nanopil should be sold in pharmacies and be available to the larger public are inadequate given the current state of the art. Similarly, in the case of the Immunosignatures, researchers at the time of the study investigated possibilities for samples to be sent to a central laboratory rather than analyzed in users' homes by a point-of-care device (see Sect. 6.4.1). This implies that scenarios of the device lying on the kitchen table should not be the focus of deliberations about the desirability of ImSg. An assessment of expectations anchors normative reflections as well as public deliberations on emerging technologies to the research practice and reduces speculations and unlikely scenarios.

Second, the analysis of the (technical) specifications of an expected artifact rules out misleading assimilations to other technologies. For example, it highlighted that the type of information about one's health condition, provided by the Immunosignatures, is epistemologically different from the one provided by genetic tests. In the case of the former, the information is contingent on a current state,

8.2 Towards Ethical Assessments of Emerging Technologies

whereas in the latter the test provides information about a disposition. As a result, the information resulting from a genetic test, that an individual belongs to a risk group, will remain with them for life, while the signature of one's immune system may change continuously (see Sect. 6.5.3). Immunosignatures share certain aspects with genetic tests, but it would be misleading to raise exactly the same ethical concerns for both tests. These types of considerations contribute to re-defining problems and issues in light of the specificity of a certain type of innovation, thereby warning against comparisons with previous cases that do not take into account relevant differences.

Third, this analysis of expectations articulates inappropriate assimilations that are often invisible to actors involved in the innovation process. For example, when I fed my analysis back to researchers from the Centre for Innovations in Medicine in Arizona, I pointed out how their expectations of Immunosignatures seemed to swing between, on the one hand a scenario in which they were used as a traditional diagnostic tool, and on the other, a scenario in which they were used as an innovative tool for personalized and preventive medicine (see Sect. 6.4.1). I demonstrated to them how these expectations imply different contexts of use, stakeholders, roles to take up and ultimately different visions of the world. After my presentation, one of the Center's co-directors acknowledged the unintentional ambiguity of their rhetoric and expressed the need for a more focused and clearly defined position about the path that they want to take. Similar observations can be made in the case of the Nanopil. In fact, Nanopil developers hold two models for how the result can be communicated to the user, by radio signal or, by releasing a blue dye. Engineers point out the technical difficulties associated with releasing a blue dye and the relative feasibility of the radio signaling. By analyzing the potential context of use of the Nanopil, I was able to explain that a decision regarding the best design should also take into account the expected context of use. In fact, the blue dye system could work for a self-test but it would prevent the Nanopil from being suitable for a national screening program for colon rectal cancer (Sect. 4.4.2). The analysis of expectations therefore has a critical distinguishing role which is an important prerequisite for the assessment of emerging technologies as it prevents an inappropriate clustering of issues. Furthermore, it also contributes to technology developers' decision-making as it shows implicit incongruences in their visions of innovations in a social context.

The plausibility assessment makes explicit the systems of values embedded in technology design and compares them with the systems of values shared by other communities of stakeholders. As discussed in the case of the Nanopil, the value of privacy and efficiency may be in tension with one another and therefore require to be balanced out. Also, empirical studies have shown that patients are more repulsed by taking a laxative than by collecting stools. Highlighting implicit conflicts between and among stakeholders' values is a first step towards a further analysis of these values. Furthermore, it may foster a pluralist democratic debate about these values at an early stage of technological development. In such a debate, diverging ethical stances and value systems are given the space for articulation. These moral disagreements are made explicit when the technology development is still ongoing and therefore can be included in the development of new technologies. For example, in the case of the Nanopil's system to communicate results, technology developers should take into account other stakeholders' considerations besides efficiency. This could imply that they have to spend more resources on the development of a technologically more difficult, yet more private, blue-dye system. This early consideration of potential moral controversies avoids polarizations later in the ethical debate, when design choices have already been made.

Every technology assessment is an evaluation and as such is based on some criteria or values. In these evaluations, moral values and norms play a role. The judgment that a desirable diagnostic technology should be cheap can refer to values of distributive justice, according to which public resources should be equally distributed in society. It can also refer to values of profit and economic return for the producer. Both judgments are based on a more or less explicitly articulated idea of "good" and in this sense they are moral values. An assessment of expectations' plausibility makes explicit the moral values and norms that are often implicit in judgments about the desirability of emerging technologies. Such increased awareness of the normative assumptions makes the debate about what is really at stake and the actors who should be involved (because they have something at stake) more straightforward. In this way, the most relevant voices and perspectives are brought to the discussion table in order to have their say.

Finally, the plausibility assessment points out the heterogeneity in expectations: different artifacts, a variety of users, diverging values and a plethora of impacts and ethical controversies. As I have argued above, at an early stage of technology development some ethical conflicts might still be latent and cover up a potentially problematic situation due to optimistic and ambiguous expectations. Such expectations might gather stakeholders together in an abstract illusion of agreement. Therefore, it is important to understand the nature of the problem explicitly presented in expectations and to critically reconstruct it, making explicit any hidden problems. By showing the diversity and heterogeneity of claims in expectations, the plausibility assessment therefore questions their apparent universality; it identifies problems and creates space for discussing alternative visions.

The natural question raised here is: to what extent does the plausibility approach contribute to the goal, outlined in Chap. 1, of addressing the normative deficit in TA? Indeed this assessment does not imply that the Nanopil or Immunosignatures are good nor does it offer a final evaluation based on moral norms and values. However, it does create the preconditions to engage in a normative discussion and evaluation. The assessment of plausibility of expectations broadens the discourse on impacts of the expected technology. According to the instrumentalist logic, a technology is designed to have a clearly defined, valuable impact. My analysis points out that emerging technologies can be expected to achieve much more than the intended effect because they mediate practices and shape (and are shaped by) standards, concepts, values and responsibilities. For example, in the case of the Nanopil, expectations carry a particular normative vision of the relationship we should have with our bodies as something we cannot trust completely (Sect. 5.4). The epistemic distance implied in the practices of monitoring ourselves with the Nanopil contributes to redefining this relationship. These types of scenarios are not predictions but rather they maintain a fictional character. They have been used to trigger researchers' moral imagination about the device. These descriptions do not suggest a specific direction to take when making decisions around the Nanopil, but they invite technology developers to think about the best system to communicate the result to the user: for example, the release of a dye in the feces instead of the radio signaling system would help the user to be more in touch with her body. Furthermore, such analysis suggests new conceptual tools that can help address problems raised by novel technologies. Such exercises of imagination about the interactions between current moral frameworks and new technologies also allow stakeholders to reflect on questions of the good life. This is a first step towards reassessing norms, re-evaluating priorities, and re-defining concepts as well as values. It helps in imagining how technologies affect our concepts, value systems and the way we conduct our lives –in short, our morality. In this dynamic process some values will be demoted in favor of others, some practices will change their meanings, some forms of life will extinguish. These are matters that are worth discussing because they concern the whole of society. In this way, not only more perspectives, but also more issues and questions become debatable and the object of deliberation. In Aristotle's words, the act of deliberation depends on the presence of alternative possibilities that we can weigh and decide upon (1954). We deliberate only on what can be different from how it is when we have a range of possibilities for our actions. The plausibility assessment presented in this book shows that multiple futures may be implicated in a single expectation. Also, it shows how some roles, responsibilities, concepts and values are contingent: they can be different from how they are now and from how they are expected to be. It makes the values and the moral concepts explicit in the discourse and includes them in the process of deliberation, rather than leaving them as an unquestioned background. As such, the plausibility assessment opens up the space and the scope of the deliberative process.

In the context of emerging technologies, a democratic society should expand the space of what is plausible for discussion and deliberation, and include more perspectives and issues. The plausibility assessment provides the conditions for realizing this goal. It does so by mediating between two extremes of 'everything is possible' on the one hand, and 'false necessity' on the other. It grounds the discussion of values and impacts in the research and social practices, excluding some scenarios and discussions as implausible. At the same time, it points out that the values inscribed in expectations of future socio-technical systems depend on specific value judgments and worldviews and are therefore contingent and questionable. Technological futures are not inevitable and since alternative futures are plausible, democratic deliberation about them should be encouraged.

8.3 Ethical Expertise? Interpreting and Intervening

The ethicist, philosopher, or social scientist[1] engaged in this type of assessment takes on the role of improving the conditions for deliberation in which social actors, who have something at stake, give and ask reasons about the desirability of emerging technologies to one another. In line with a pragmatist approach, the issue is not about offering a final universal normative prescription, but is instead about the recognition of pragmatic sites for ethical discussions. To this aim, an *analytical* and an *interventionist* phase are necessary. The *analytical* phase consists of assessing the plausibility of expectations about the artifact; its use and its desirability. Such phase requires practical involvement; it is not an activity for "armchair" philosophers. Rather, it requires fieldwork and mingling with stakeholders in the "real world".[2] This fieldwork provides the grounds for re-describing the expectations of emerging technologies. The aim of these re-descriptions is to elicit the morality that is sometimes implicit in the expectations of the artifact and its use. This is important in order to uncover the heterogeneity in these expectations and to identify alternatives within current visions of futures. The *interventionist* phase consists of feeding this analysis back to stakeholders. In order to do this, specific tools have to be designed. The scenario workshops with technology developers, presented in Chap. 7, are examples of these tools. Multi-stakeholder workshops or public debates are other examples of how this analysis can be fed back into society. Generally speaking, these tools need to foster deliberation rather than having a report-like form.

The plausibility analysis has an "epistemological" rather than a "prescriptive" role because it aims to improve the conditions for developing knowledge about technologies rather than offering a judgment on whether they are good or bad. Within the pragmatist approach, the expertise of the ethicist does not consist of judging what is good or bad about emerging technologies. Although the "ethicists" may have a "procedural" expertise of gathering, analyzing and interpreting the moral conflicts and normative visions, they do not have a preferred "moral competence" that legitimizes their position over other stakeholders. The ethicist can help improve the debate, guaranteeing that some moral issues are addressed without offering prescriptive solutions, on the one hand, and without leaving the deliberative process totally to stakeholders' unexplored preferences, on the other. It is not simply a matter of collecting stakeholders' preferences, but is rather about creating

---

[1] When I am speaking of the "ethicist" I am not referring to a person with a specific disciplinary background, but a "humanist" engagement in the type of work described so far of exploring normative visions and creating preconditions for normative deliberation (See also Boenink 2013).

[2] The distinction between ethical debates and "real world" ethics is the starting point of the EU funded project "DEEPEN". Its declared aim is to reach an "integrated understanding of the ethical challenges posed by emerging nanotechnologies in real world circumstances" (see http://www.geography.dur.ac.uk/projects/deepen/Home/tabid/1871/Default.aspx) for an insightful discussion of this topic see Shelley-Egan 2011.

conditions in order to critically understand their assumptions.[3] Furthermore, I pointed out that the moral content does not only exist in stakeholders' visions but also in technological artifacts and their design. This is particularly important when technologies are still emerging and design choices are still open for discussion.

What is "ethical" in such an approach that does not offer a normative answer to the question of the desirability of emerging technologies? This approach promotes conclusions that might work for the moment, without reaching the ideal endpoint of absolute "moral correctness". The contingent conditions of space, time and cultures are taken seriously into account within this perspective. Despite the fact that moral norms have to be shared for functional cooperation in any society, their contingency is, and should be made, clear and legitimate. It reveals problematic situations that need to be addressed because they are characterized by moral ambiguities, uncertainties and challenges. This assessment does not offer solutions, but fosters a reconsideration, from within, of the accepted normative framework.

8.4 Open Questions

Several questions remain open for further exploration. The first set of questions concerns the meaning of "democratic deliberation" for technology assessments of emerging technologies. Throughout the book I talked about "democratic deliberation" in several instances. I explained in what sense deliberation around emerging technologies should be democratic. This was intended as a broadening of the issues that included moral concerns and normative discussions concerning the acceptability or desirability of a certain technology, as well as the inclusion of a broad and diverse range of stakeholders. I also explained that such a deliberation happens in a context where arguments are weighted and reasons are given, asked and discussed. But where does such democratic deliberation take place? Who are the participants? Are exercises of public engagement, like consensus conferences, an instance of this? Or are CTA-like workshops a better example? Furthermore, how can the workshops described in Chap. 7 be considered as an instance of democratic deliberative exercises if they are characterized by such a homogenous group of participants?

Many initiatives trying to engage the broader public in deliberative exercises have proven to reiterate power relationships in their design, to miss the opportunity to explore normative issues or to be unable to influence policy decision- making (Genus and Coles 2005, see also Chap. 1). I have discussed the limitations and opportunities of workshops with scientists in Chap. 7 and revealed how they are an attempt to address the issue of exploring moral aspects and are designed to engage technology developers in an early discussion on the moral good of their research. Although the workshops did not include a broad set of stakeholders, some actors' perspectives were included in the scenarios. How to turn these discussions into concrete decisions at the lab level or at the policy level? This remains an open question

---

[3] See for example the third point made in Sect. 8.2.

that this book has not directly addressed. Indeed the current calls for Responsible Research and Innovation at the European and national level,[4] suggest that the systematic inclusion of normative discussions at the research and innovation level is a concern of policy makers (Grunwald 2011). However, it remains to be explored how the type of tools discussed in this study may be integrated into a more systematic attempt of RRI.

A second group of questions relates to cultures of plausibility. As was pointed out in Chap. 2, expectations are more than statements about future states of affairs. They have a driving and strategic role, they do things. The discussion of the literature on visions has emphasized how this driving force has a normative dimension, how visions of desirable worlds move imaginations (together with resources and consensus). Visions are expressed not only in discourses, as verbally expressed scenarios. They are also objects and practices. The plausibility analysis, in fact, has not only focused on discursive judgments of plausibility. Images, models, laboratory experimental set-ups and research directions were analyzed in their unfolding in technology developers' practices and actions. One aspect that would need to be further explored is how these cultures of plausibility are characterized. I have shown that what is plausible for researchers may not be so for other stakeholders and I have pointed out how these frameworks of plausibility depend on both the actors' beliefs and values. A more systematic study could unpack the way in which actors with slightly different knowledges and values (say researchers coming from different disciplinary background) regard the plausibility of the same vision.

Thirdly, this study does not address the issue of whether the Nanopil or Immunosignatures are good or bad technologies. It has not offered prescriptions on whether they should be pursued or not. Rather, it has shown how to create conditions for a contextual normative discussion that can happen in a specific setting of deliberation among different stakeholders, policy making or among scientists in the lab. What remains to be addressed is how we can create such spaces for local assessments, in which stakeholders do not take a passive role, as was seen in the workshops described in Chap. 7, but instead engage in normatively-directed deliberations. It is also remarkable that some of the issues raised cannot be discussed by scientists alone. For example, the question of whether the Nanopil should be developed with a blue dye system or an electronic signal system cannot be left to the deliberation of only scientists. Technology developers may say that it is very hard to install a micropump in such a small mechanism. However, a broader discussion with other users and policy makers may arrive at the conclusion that the pill should not produce data that could be hacked and would work better as a thermometer. The issue is how to ensure that a discussion about technical possibility is conducted among different groups? Of course it is hard to imagine something like this in the current system, but again this is something to investigate in the context of the distribution of responsibilities and tasks among the broad range of stakeholders involved, which goes

---

[4] See for example article 5 of the Regulation REGULATION (EU) No 1291/2013, establishing Horizon 2020, available at http://eur-lex.europa.eu/LexUriServ/LexUriServ.do?uri=OJ:L:2013:347:0104:0173:EN:PDF.

beyond the development of a technological artifact. Another example concerns the fear that by trusting technologies, patients and citizens lose epistemic responsibility towards listening and learning from their bodies (Chap. 5). This is, of course, an issue that does not only pertain to the technology at stake, in this case the Nanopil, and its developers, but also concerns a broader range of actors. For example, the patient organization or the Ministry of Health may decide to foster a campaign to raise people's awareness and educate them that when using the Nanopil they should learn also how (as one of my interviewees said) to "feel their gut". How to ensure that different actors and alternatives contribute to a vision of a desirable society that is not only technology driven and focused?

Fourthly, there is the aspect of the "ethics of promising". If any at all, the prescriptive purport of this study is at the meta-level and concerns the way technologies should be promised. For example, from the plausibility analysis it emerges that it would be immoral to discuss what type of society would be the one in which the Nanopil sits on grocery store shelves. Within our current state of knowledge, based on ongoing research, the Nanopil does not have a shelf life. It may have one in the future, but this is a future world we do not know much about at this stage (Nordmann 2013) and we should carefully avoid any sterile speculation about it. Of course when engaged in public discussions, technology developers are not the only ones setting the debates: ethicists or "the broader public" often discuss science fictionary scenarios in science cafes. However, they should be able to recognize when ethical questions are out of place or technically implausible. As such, the plausibility assessment is useful for highlighting such implausible cases. It remains, however, to be seen who is responsible in promising technologies and also, how to identify specific contexts and practices in which different actors hold different roles and responsibilities for promising. Indeed, what would such an "ethics of promising" look like?

References

Aristotle. 1954. In *Rhetoric*, ed. W. Rhys Roberts. New York: Cambridge Univ Pr.
Boenink, Marianne. 2013. The multiple practices of doing ethics in the lab. In *Ethics on the laboratory floor*, ed. Simone van der Burg and Tsjalling Swierstra. Basingstoke: Palgrave Macmillan.
Genus, Audley, and Anne-Marie Coles. 2005. On constructive technology assessment and limitations on public participation in technology assessment. *Technology Analysis & Strategic Management* 17(4): 433–443. Routledge.
Grunwald, Armin. 2010. From speculative nanoethics to explorative philosophy of nanotechnology. *NanoEthics* 4(2): 91–101.
Grunwald, Armin. 2011. Responsible innovation: Bringing together technology assessment, applied ethics, and STS research, November. IET. http://run.unl.pt/handle/10362/7944
Keulartz, J., M. Schermer, M. Korthals, and T. Swierstra. 2002. *Pragmatist ethics for a technological culture*. Dordrecht/Boston: Kluwer Academic.
Nordmann, Alfred. 2007. If and then: A critique of speculative nanoethics. *NanoEthics* 1(1): 31–46.
Nordmann, Alfred. 2013. (Im)Plausibility2. *International Journal of Foresight and Innovation Policy* 9(2-3-4). Inderscience Publishers: 125–132.

Nordmann, Alfred, and Arie Rip. 2009. Mind the gap revisited. *Nature Nanotechnology* 4(5). Nature Publishing Group: 273–274.

Shelley Egan, Clare. 2011. *Ethics in practice: Responding to an evolving problematic situation of nanotechnology in society*. Enschede: University of Twente.

Swierstra, Tsjalling, and Hedwig Molder. 2012. Risk and soft impacts. In *Handbook of risk theory SE – 42*, ed. Sabine Roeser, Rafaela Hillerbrand, Per Sandin, and Martin Peterson, 1049–1066. Dordrecht: Springer.

Thompson, Paul B. 2002. Pragmatism, discourse ethics and occasional philosophy. In *Pragmatist ethics for a technological culture SE – 16*, vol. 3, ed. Jozef Keulartz, Michiel Korthals, Maartje Schermer, and Tsjalling Swierstra, 199–216. Dordrecht: Springer. The International Library of Environmental, Agricultural and Food Ethics.

If you have any concerns about our products,
you can contact us on
ProductSafety@springernature.com

In case Publisher is established outside the EU,
the EU authorized representative is:
**Springer Nature Customer Service Center GmbH
Europaplatz 3, 69115 Heidelberg, Germany**

Printed by Libri Plureos GmbH
in Hamburg, Germany